The Open University

Block 3

Creativity and concept design

Prepared on behalf of the Course Team by Robin Roy

T211 Design and designing

This publication forms part of an Open University course T211. *Design and designing*. Details of this and other Open University courses can be obtained from the Student Registration and Enquiry Service, The Open University, PO Box 197, Milton Keynes MK7 6BJ, United Kingdom: tel. +44 (0)845 300 60 90, email general-enquiries@open.ac.uk

Alternatively, you may visit the Open University website at http://www.open.ac.uk where you can learn more about the wide range of courses and packs offered at all levels by The Open University.

To purchase a selection of Open University course materials visit http://www.ouw.co.uk, or contact Open University Worldwide, Walton Hall, Milton Keynes MK7 6AA, United Kingdom for a brochure. tel. +44 (0)1908 858793; fax +44 (0)1908 858787; email ouw-customer-services@open.ac.uk

The Open University
Walton Hall, Milton Keynes
MK7 6AA

First published 2004. Second edition 2009.

Edited and designed by The Open University.

Printed in the United Kingdom by the Charlesworth Group, Wakefield

ISBN 978 0 7492 5434 6

2.1

Contents

Introduction 5

Aims and learning outcomes 6

Study Chart 8

Part 1 Theory

1 Creativity in design and innovation 10
- 1.1 Introduction 10
- 1.2 What is creativity? 10
- 1.3 Creativity and concept design 13
- 1.4 Design thinking 28

2 The creative process 35
- 2.1 Bicycles – historical examples of creativity 35
- 2.2 Theories of creativity 38
- 2.3 Creativity and innovation 48

3 Creative designing – individual approaches 52
- 3.1 Bicycles – contemporary examples of creativity 52
- 3.2 Principles of creative designing 68

4 Creative designing – systematic and team approaches 75
- 4.1 Mountain bike 'Splashguard' 75
- 4.2 Bicycle backpack carrier 81
- 4.3 Creative design teams 83

Part 2 Practice

5 Practical creativity for design 88
- 5.1 Creative designing requirements 88
- 5.2 Experienced designers' approaches to creativity 90
- 5.3 Creative problem solving techniques 101

6 Creativity is not enough 109
- 6.1 Obstacles to innovation 109
- 6.2 Obstacles to diffusion 111
- 6.3 Three examples of innovative products 115
- 6.4 Conclusion 125

Answers to self-assessment questions 127

References 133

Acknowledgements 135

Course Team 136

Case studies and examples

Concept design of a car horn 29
The 'Rover' bicycle 35
The pneumatic tyre 37
The 'Cyclone' cleaner 48
The Moulton small wheel bicycle 52
The 'Brompton' folding bicycle 60
The 'Strida' folding bicycle 63
Mountain bike 'Splashguard' 75
Bicycle backpack carrier 81
The Freeplay clockwork radio 115
The Sinclair 'C5' 121
The 'Segway' human transporter 124

Exercises

1 Ideas about creativity 10
2 Elements of creativity 18
3 Types of concept design 23
4 Ideas, concepts, invention and technical innovation 27
5 Alarm clock concepts 32
6 Lessons of the Moulton bicycle 55
7 Lessons of the Brompton bicycle 63
8 Elements of creative designing 69
9 Identifying functions 77
10 Sketching concept designs 80
11 Creative design requirements 88
12 Information sources 92
13 Thinking in terms of function/performance 94
14 Associative thinking 95
15 Evolving technologies and materials 96
16 Socio-economic trends 97
17 LCA for a clockwork radio 119
18 Successful innovation 124

Introduction

Block 3 concentrates on the creative aspects of designing. The block focuses on creative thinking and working in the vital early phase of designing during which, in response to a design brief, need or problem, the basic idea for a new product is conceived and given physical form in an initial concept design. However, the block emphasises that creativity is required throughout the product development process – from the perception of product and user problems, or of market and technological opportunities, to the resolution of detailed design and manufacturing issues.

The block will help you to develop your creative design abilities by learning about the theory and practice of how highly creative individuals and teams conceive new product ideas and develop them into concept designs or inventions. Case studies of the design of bicycles and their components provide a common theme throughout the block, supplemented with examples from other fields presented in the block text and on the course DVD.

You are taught creative thinking and concept designing skills to prepare for the Block 3 TMA. These include gathering information, associative thinking, generating ideas as an individual and in groups, and the use of sketching and 3D modelling. You are also provided with selected creative problem solving techniques, such as mind mapping, via the block text and interactive exercises and creativity software on the DVD.

You will learn that being creative is not enough to ensure that a new product succeeds on the market, meets user needs or environmental requirements. It is necessary to integrate creative designing and inventing with market, technical, materials and manufacturing constraints and environmental impacts.

Aims and learning outcomes

Aims

1 To provide an understanding that creativity is required throughout new product development, especially at the early concept design phase at which the operating principles and forms of designs are generated and selected.

2 To provide an understanding of the creative thinking processes and the approaches and methods used by experienced designers and design teams for exploring design problems and for generating and selecting ideas and concept designs.

3 To teach selected systematic design methods and creative problem solving techniques to enhance individual and group creativity, especially of learner designers.

4 To enable you to choose appropriate approaches and techniques, apply these to generate creative solutions to simple concept design problems, and to select one or more concepts for further development.

5 To help you to apply the creative thinking skills learned in the context of concept design throughout the product development process, and to problem solving more generally.

6 To provide an understanding that creativity is not enough to ensure that an innovative product is successful on the market, meets users' needs, or reduces environmental impacts.

Learning outcomes

As a result of studying Block 3 you should have achieved the following learning outcomes:

1 Knowledge and understanding

You should be able to demonstrate knowledge and understanding of:

1.1 the importance, and the limitations, of creativity in the development process for innovative products;

1.2 the meaning and main theories of creativity and the thinking processes involved, especially in invention and concept design;

1.3 the factors, sources and triggers that may help or inspire individual designers and design teams to produce creative ideas;

1.4 the role of 2D sketching and drawing, 3D physical models and computer-based modelling in invention and concept design;

1.5 the factors other than creativity that influence the development of a successful innovative product (e.g. cost, competitive advantage, fashion, ease of manufacture, team working and environmental impacts).

2 Cognitive (thinking) skills

You should be able to:

2.1 identify creativity in the designing processes and concept designs of others;

2.2 apply, to simple concept design problems, heuristic approaches used by experienced designers and design teams, plus selected systematic techniques that can aid learner designers;

2.3 generate ideas and concept designs using computer software to aid creative thinking;

2.4 evaluate your own creative ideas and concepts, and those of others.

3 Key skills

You should be able to:

3.1 apply skills of creativity including the exploration of problems and the generation and selection of ideas;

3.2 select suitable approaches and techniques to help you to 'flex' (move) away from your usual mode of thinking and thereby become more creative;

3.3 consider and evaluate environmental impacts when generating and selecting ideas and solution concepts;

3.4 apply appropriate working styles, e.g. individual, group or team, to tackle problems requiring creativity.

4 Practical skills

You should have gained:

4.1 the skills necessary to use computer and Internet resources to obtain relevant information for creative problem solving;

4.2 modelling skills (e.g. sketching) to apply in the conception, development and communication of ideas and concept designs.

Study Chart

Section	Main text	Workbook 1	DVD	Block 3 Supplement
Part 1 Theory				
1	Creativity in design and innovation	Section 1 (revision)		Study Guide
2	The creative process		James Dyson: designer and innovator	DVD and computing notes
3	Creative designing – individual approaches		Mark Sanders: designer and engineer	DVD and computing notes
4	Creative designing – systematic and team approaches		IDEO: creative design teams Philips: creative product design	DVD and computing notes
Part 2 Practice				
5	Practical creativity for design	Sections 4–5 (revision)	New designers Creative problem solving techniques	DVD and computing notes
6	Creativity is not enough	Section 6	MET matrix technique Video of *Modelling Workbook* Activities 11–15	DVD and computing notes

Support notes for Block 3 TMA appear in the *Block 3 Supplement*. You should refer to these as you prepare to answer TMA 03.

Part 1
Theory

1 Creativity in design and innovation

1.1 Introduction

In this block you will learn that creativity is required *throughout* the processes of new product development but especially during the concept design phase. As you saw in Block 2, the designing process proper often starts with the issue of a brief or specification. But, as this block shows, the starting point may not be a formal – or even an informal – brief. Creative ideas and concept designs often emerge from a problem or opportunity perceived by the designer.

1.2 What is creativity?

Since this block is about being creative, let's look first at what you understand by creativity.

Record your answers to this and subsequent exercises in your workfile.

Exercise 1 Ideas about creativity

(a) Name at least three people whom you consider *exceptionally* creative in a particular field or fields (e.g. art, design, engineering, science, music).

(b) List at least three products or works that you consider to be the outcome of the activities of exceptionally creative people. These could be the work of the individuals you named in part (a).

(c) List at least three products that you consider the results of 'ordinary' or 'everyday' creative activity.

(d) List at least three products that you don't consider the result of creative activity.

(e) What distinguishes the products or works of exceptionally creative processes and/or people?

(f) What is the difference between such exceptionally creative works and the products of ordinary or everyday creative activity?

(g) What are the characteristics of the products of non-creative activities?

Spend 10–15 minutes on this question before looking at my response below.

My response

(a) My list of exceptionally creative individuals (and their main field of creative activity) includes:

- Leonardo da Vinci (painting, sculpture, anatomy, invention);
- Thomas Edison (electrical inventions);
- Isambard Kingdom Brunel (civil and mechanical engineering);
- Pablo Picasso (painting);
- Douglas Adams (science fiction);
- Mary Quant (fashion clothing);

- Philippe Starck (furniture, household goods, architecture);
- John Lennon and Paul McCartney (popular music);
- Alec Issigonis (motor cars);
- Steve Jobs (personal computers);
- Tim Berners-Lee (computing, the Internet);
- Stephen Hawking (theoretical physics);
- Daniel Libeskind (architecture).

Many of these examples betray my age! You may have come up with some more recent ones.

(b) The products or works of these exceptionally creative individuals include:

- paintings such as *The Last Supper* and the *Mona Lisa*; understanding of human anatomy and physiology; inventions for the parachute (Figure 1), helicopter, submarine, etc. (Leonardo da Vinci, 1452–1519);
- the phonograph; the electric incandescent lamp; (Edison, 1877, 1880);
- the Clifton suspension bridge, Bristol; the broad-gauge Great Western Railway; *The Great Eastern* and other steamships (Brunel, 1806–1859);
- *Les Demoiselles d'Avignon* (painting by Picasso, 1907 showing for the first time the non-representative human figures that would characterise much 20th century art);
- *The Hitchhiker's Guide to the Galaxy* (radio scripts and book by Douglas Adams, 1978–9);
- the mini-skirt (Mary Quant, also André Courrèges, 1964);
- the 'Juicy Salif' lemon squeezer (Philippe Starck, 1990);
- *Sergeant Pepper's Lonely Hearts Club Band* (recorded songs and music by The Beatles, 1967);
- the Austin/Morris Mini car (Alec Issigonis, 1959);
- the Apple Macintosh personal computer (Steve Jobs and others, 1984);
- the World Wide Web (programs for a global computer network of linked hypertext documents, Tim Berners-Lee, 1989–90);
- *A Brief History of Time* (non-specialist book describing Stephen Hawking's theoretical work on space and time, 1988);
- The Jewish Museum, Berlin and Imperial War Museum North, Manchester; winning design for rebuilding the World Trade Centre, New York (architecture by Daniel Libeskind, 2001–3).

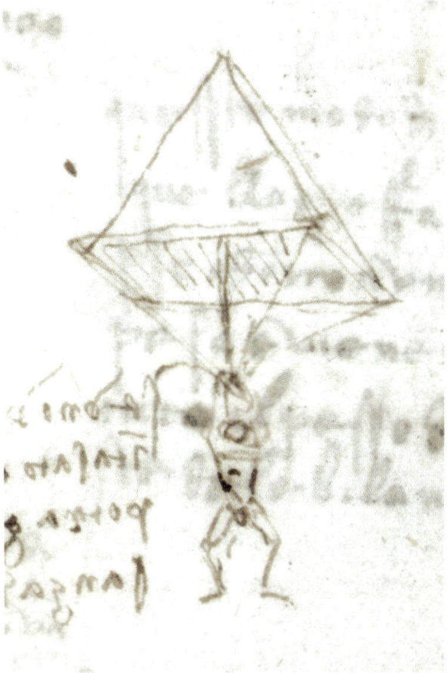

Figure 1 Leonardo da Vinci's invention and design of a parachute; intended to allow a person to jump from any height without risking injury

(c) Products resulting from 'ordinary' or 'everyday' creative activities or people:

- a meal made, or modified, to my own recipe;
- the colour scheme and furnishing of a room;
- a design for a garden;
- the pages of a website;
- a craft-made clay pot;
- furniture from a mass market manufacturer;
- a speculatively built house.

(d) Some products that I consider are *not* the result of creative activity:

- 'painting by numbers';
- a completed jigsaw puzzle;
- design of a steel shaft to transmit a given amount of power from an electric motor;
- choosing a gearwheel for a new product from a catalogue of standard engineering components.

(e) What distinguishes the products or works of exceptionally creative processes and/or people?

novelty
Being new or different from what was previously known or usual.

originality
Being the first example of something; not derived from anything that has appeared before.

value
Having importance, significance, worth, utility, or desirability.

First, they display exceptional novelty or originality, representing an (apparent) break or discontinuity from what has gone before, anywhere in the world.

Second, the exceptionally creative works are (at least eventually) generally accepted as being of great value to many people in the world. They have a major influence on the work of others and/or may be of importance in shaping whole cultures or technologies. They may provide the foundation of new businesses or industries, or even affect the performance of economies and nations.

(f) What is the difference between such exceptionally creative works and the products of ordinary or everyday creative activity?

Products of 'ordinary' creativity may involve their creator in producing what is *for them* a novel or original work, but are not generally considered as representing a significant change or discontinuity from what already exists.

Products of 'ordinary' creativity may be of value to their creator, or to a particular group of people, but are not usually of special importance to the world at large.

Weisberg (1993) in his book *Creativity: Beyond the myth of genius* emphasises that people who produce exceptionally creative works are not 'genuises'. They use the same thought processes as we do when producing 'ordinary' creative work. When such people produce exceptionally creative works it is their knowledge, motivation and dedication that makes the difference – not different thought processes. This means everyone can be creative – you don't have to be exceptional.

(g) What are the characteristics of the products of 'non-creative' activities?

No novelty is involved – the object, solution or design is fixed in advance or is a copy, or at most a slight variant, of existing products.

Weisberg also argues that novel work that is the result of an accidental, random or irrational process *does not* involve creativity. Creative activity involves a *goal* or intention. Thus paint accidentally spilled on a canvas would not involve creativity unless its creator recognised the potential of this method to produce 'worthwhile' art.

Creativity
A *goal-directed* human activity that results in *novel* or *original* work that is perceived to be of *value* to individuals or society.

Thus two definitions of creativity are:

> Creativity is … a novel and appropriate, correct, useful or valuable response to the task at hand.
>
> (Amabile, 1996, p. 35)

> Creativity is imaginative activity fashioned so as to yield an outcome that is of value as well as original.
>
> (National Advisory Committee on Creative and Cultural Education, 1999, p. 29)

There is clearly a *spectrum* of originality, novelty and value – from breakthroughs of significance to the whole world (e.g. the first powered aircraft, the Internet) to the products of everyday creativity of value mainly to their creator and immediate group (e.g. a flower arrangement, a room interior). Even so, how creative something is considered to be is a matter of judgement and thus will vary in time and place.

In this course we are mainly concerned with designing products of intermediate novelty and value along this spectrum – from bicycles to buildings. You'll see that individuals or teams who produce such creative works tend to share certain characteristics, and ways of thinking and acting. Many of these can be learned. So, although *everyone* has creative ability, they can also enhance their skills by learning from those who regularly produce creative work.

In this block I shall provide you with materials that should enhance your creative design ability, by teaching you how experienced designers work and by providing you with some creative problem solving techniques that can help 'learner' designers.

1.3 Creativity and concept design

1.3.1 Elements of creativity

You should now have some idea of what is meant by creativity, but what does creative activity actually involve? In particular what's involved in the concept stage of designing? To begin to understand the nature of the creative process in concept design and product development, let's consider some examples.

First, a historical example. Here's an account of how a veterinary surgeon, John Boyd Dunlop, the second and successful inventor of the pneumatic tyre, came up with his idea in the 1880s (the first inventor Robert Thompson did not make a success of the original pneumatic tyre):

> Dunlop's mind, like Thompson's and many others before him, had been considering the problem of vehicle vibration. For a period of over twenty years Dunlop had considered, and then abandoned as impractical, various forms of spring wheels, including the idea of a flexible steel band supported by rollers. Eventually he thought of the idea of a wheel with an air-filled tyre, which he envisaged as being made of cloth, rubber and wood. Once the idea of an air-filled tyre had occurred to Dunlop, he rapidly proceeded to conduct a rough experiment. For the experiment Dunlop used a solid disc of wood to which he nailed a pocket of linen that enclosed an inflated rubber tube. He compared the rolling properties of this wheel with a solid rubber-tyred wheel, with promising results: 'I first threw the solid tyred wheel along the yard, but it did not go the full length of it. Then I threw the air tyred disc. It went the whole length of the yard and rebounded with considerable force off the gate.'
>
> (Dunlop, 1922, p. 11)

Second, Alex Moulton, a creative engineering designer (whose best-known work, the design of small-wheel bicycles is described later in this block), notes what was involved in developing the innovative 'Hydrolastic' suspension system used on the BMC Mini and the BMC 1100/1300 series of cars in the 1960s (Figure 2):

> These things do not appear on the stage by a flash of insight – important as this is – but from years of experiment, observation and consequent modifications. Above all it involves countless hours of thought and discussion with the intimate few …
>
> (Moulton, 1979, p. 39)

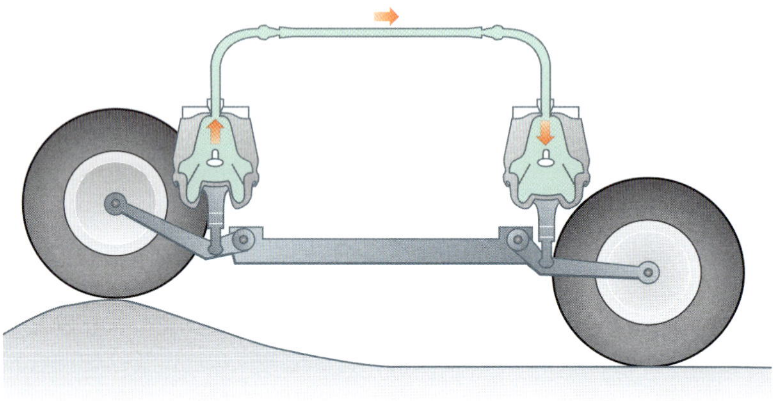

Figure 2 The 'Hydrolastic' suspension system

This uses fluid connection between front and rear conical rubber springs on each side of a car. When the front wheel hits a bump it rises and thus displaces fluid from the front spring which enters the rear spring. This lowers the rear wheel, which helps keep the rear of the car level and smoothes the ride. The concept was tested in 1955 by Moulton and Alec Issigonis on a prototype Alvis car. Interconnected fluid car suspension systems were developed simultaneously and introduced first in France by Andre Citroën (Nahum, 1988, p. 38; p. 68).

Third, here's what an engineer and industrial designer (who appears on the DVD sequence, 'Mark Sanders: designer and engineer') said about how he conceived his novel 'Strida' folding bicycle (Figure 3):

'A long time was spent thinking about the problem and jotting down ideas at random.'

(I've repeated some parts of the quotes like this in the margin throughout the text, to remind you of different designer's creative approaches.)

> A long time was spent thinking about the problem and jotting down ideas at random … The competitors all appeared to have been designed to fold as small as possible in all directions. This was questioned: why not design the folded form long and thin like a walking stick? This decision was confirmed by considering one of the most successful products of recent years, the folding baby buggy, which folds to form a long thin stick with wheels at one end … Having decided on this folded form, the structure of the bicycle was concentrated upon.
>
> (Sanders, 1985, Section 2.2)

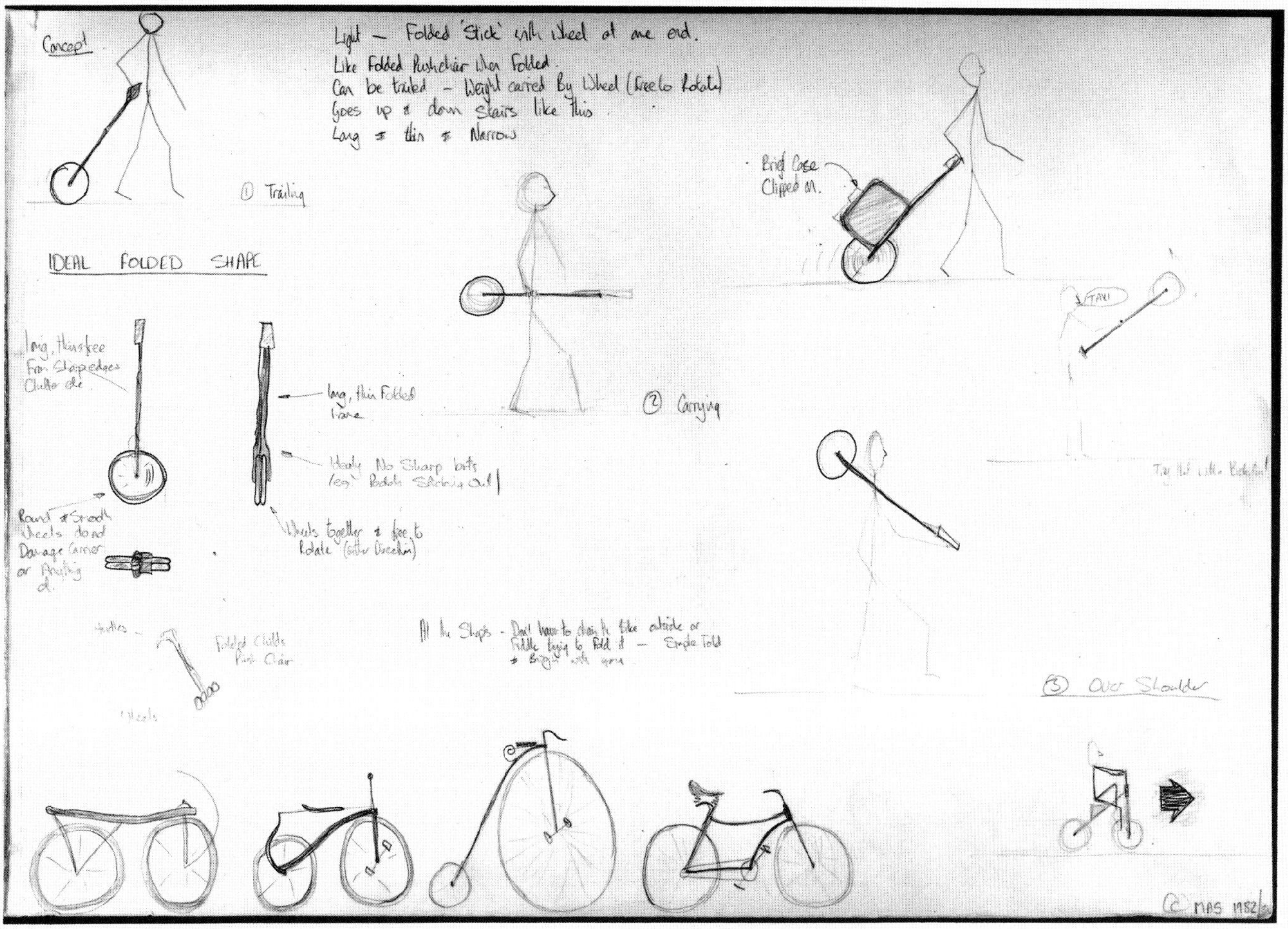

Figure 3 Page from Mark Sanders' first bicycle project book showing the basic concept of a bike which folds into a stick with wheels at one end

Fourth, here is what John Galliano, Artistic Director for Christian Dior, said in an interview about the value of the research trips he makes to inspire his collections of women's high fashion clothing:

'We sketch, buy postcards, go to flea markets. Anything to fire our imaginations.'

> First of all we put everything we've collected on the trip into what we call 'the bible' – everything, even metro tickets or weird Polaroids of us in Russia having to put these plastic duster covers over our boots … We sketch, buy postcards, go to flea markets. Anything to fire our imaginations and we bring it all back and stick it in the book. It shows how our minds are working and everyone is inspired by it. … I have done a book for every collection that I have ever produced.
>
> (Tyrell, 2002, pp. 18–19)

Fifth, below is an account of how Philippe Starck, the renowned French designer, conceived the 'Juicy Salif' lemon squeezer (Figure 4) as part of a commission to design a range of kitchenware from Italian housegoods manufacturer Alessi:

> After meeting Alessi he takes a quick break on the Italian island of Capraia. In the evening he ambles along to a local restaurant … As he sits waiting for his food to arrive he turns the problem of the lemon squeezer over in his head. He thinks about a conventional squeezer and sketches the form on the restaurant place mat. … the food has arrived – baby squid … just as he is about to put it into his mouth he stops, looks at it,

'His sketches … remind him of the comics he used to read. … Like one of the spaceships he used to think about jetting up to space in.'

and realises that this is the solution. … He starts to sketch … First he tries to make a conventional lemon squeezer out of a squid, but then he realises it won't really work. The squid begins to evolve … into something with legs, but he doesn't like it. It seems to be dragging … He keeps going, eating while he sketches. His sketches abstractly remind him of the old comics he used to read. … Things begin to gel in his mind, and from the dragging creature emerges a lighter three legged form. Like one of the spaceships he used to think about jetting up to space in. He likes the form. … The next morning he phones Alessi 'I've got a lemon squeezer for you' he teases.

(Lloyd and Snelders, 2001, pp. 244–5).

(a) Table mat showing Philippe Starck's original sketches for the lemon squeezer

Note the sketches generally seem to follow an anti-clockwise path from the bottom right.

(b) The production version of the lemon squeezer

(There is a 3D animation of the Juicy Salif on the DVD.)

Figure 4 The 'Juicy Salif' lemon squeezer and its genesis

Sixth, here's what Martin Hunt, an award-winning ceramics designer and partner in a firm of highly successful ceramics design consultants, says about what he needs to produce his creative designs:

> I need two kinds of input before I produce new ideas. I need the information that comes from discussions we have internally or with our clients. We go into some depth about the market they are aiming at [and] on their manufacturing and technical facilities. [...]
>
> Having briefed myself about what I am trying to achieve, I need the input that will get my mind going quickly. ... I love to look at magazines. It is important that they're only tangentially to do with pottery. If I sit down with *Casa Vogue* [a fashionable interiors magazine] ... I get ... a sense of what people find enjoyable and fun ...
>
> ... it never fails. Somehow an idea comes into my mind ... like an out of focus colour slide. Say the project is a teapot, the image in my mind has perhaps ... a seductive combination of colours [...] You can't copy it because it's out of focus ... so once I've got a quick sketch that's got some snap to it, I quickly get into model-making. When you're making something, you see things happening you haven't thought of.
>
> (Hunt, 1991, pp. 174–5)

'Once I've got a quick sketch that's got some snap to it, I quickly get into model-making. When you're making something, you see things happening you haven't thought of.'

Finally, James Dyson talking about his approach to invention and design. He talks on the DVD sequence, 'James Dyson: designer and innovator' about the creation of his novel 'Ballbarrow' wheelbarrow and 'Cyclone' cleaner (Figure 5):

> I am a great believer in two things. Firstly ... select a finite area and it is amazing how quickly you can become expert in it. Secondly, I believe in painstaking and logical testing. In every one of our projects, the patentable and inventive parts have been happened upon during the course of practical testing.
>
> (Dyson, 1987, p. 17)

'Select a finite area and it is amazing how quickly you can become expert in it.'

(a) How the Cyclone cleaner works

A fan sucks in air and dirt through the cleaning head or hose (small black arrows). This dirty air (black) enters the first cyclone at the top of the cylinder and swirls downward like an inverted cyclonic storm or tornado, throwing dirt to the side from where it falls to the bottom. Partly cleaned air (grey) enters the second stage cyclone where fine dust is thrown to the sides and also falls down. Clean air (white) is expelled through the fan.

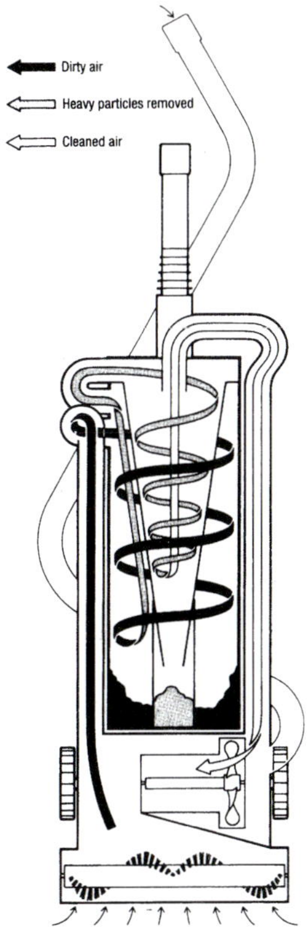

(b) Dyson DC01 cleaner, launched in the UK in 1993

Figure 5 The Dyson Cyclone cleaner and how it works

Exercise 2 Elements of creativity

Within the above quotations are many clues as to the nature of the creative processes underlying the design of novel products. What key elements of the creative process can you identify from them?

Spend 10 minutes rereading the quotations in Section 1.3.1 and making some notes in your workfile before looking at my response below. Read my response carefully as it contains a lot of points. But don't worry if you only identified a few elements. These and other elements of creativity will occur repeatedly as you study the block.

My response

1. Dunlop, Moulton and Sanders specifically mention spending *much time and effort thinking about a problem and examining different ideas* before the basic idea for a solution arose in their mind. Developing the idea also required focused effort over a long period.
2. Dunlop, Moulton, Hunt and Dyson all stress the importance of making a *physical model, mock-up or prototype* in order to test their basic ideas. If proven, the idea is gradually evolved through further models and prototypes.
3. Moulton and Dyson note the necessity for *specialist knowledge* in the field of the object being designed in order to develop it beyond just an idea.
4. Sanders and Starck say how their basic concept was *inspired or triggered by some other object*, from a folding baby buggy to a baby squid. Galliano notes that he collects a variety of photographs and objects in order to inspire his team's clothing designs, while Hunt looks at magazines to create images in his mind that lead to new ceramics designs.
5. Starck and Sanders make extensive use of *sketching* to explore the germ of an idea and evolve it into a concept design.
6. Moulton, Galliano and Hunt mention the value of *teamwork* and *discussion* in creating and developing concept designs.
7. Successful, creative individuals, like Hunt and Dyson, are not just good at producing novel ideas and products. They can also spot actual or potential *needs that are within their capacity to fulfil*. And Hunt specifically notes the importance of identifying the *market* and *manufacturing constraints* within which he has to create his designs.

You may have identified only one or two from my list, or noticed other elements of creativity from the examples. In fact each example, when examined in greater depth, contains many more elements. As you work through this block I shall be looking at these and other examples to help you understand the creative process more fully.

1.3.2 The concept phase of designing

The designing of products typically proceeds through a number of phases as the object concerned is gradually defined more and more precisely until it is described in sufficient detail for it to be made.

To refresh your memory, Box 1 reproduces the product development process as outlined in Section 9.4 of Block 1:

Box 1 Phases in designing

Task or problem clarification

This is the stage during which the *brief* is elaborated from a set of broad objectives into a more detailed list of market and technical requirements. Identifying a *specification* for the task usually involves market research and sometimes preliminary technical feasibility studies.

For example, if a manufacturer of powered gardening tools sees an opportunity to develop a new form of hedge cutter, at this stage it will be necessary to identify the potential market, the performance requirements, target costs, production constraints etc. The integration of these becomes part of the design problem.

Generating a basic concept

This is the stage during which alternative solution principles and concept designs for resolving the problem are generated and selected for further development. It is generally considered the highest level of design ... because if the concept design is wrong no amount of subsequent development ... will make it right. [...]

Using our example of a new hedge trimmer, a manufacturer may consider different principles for cutting hedges – reciprocating knives, rotating cutters, rotating wire, high pressure water jet, freezing, burning (!) – and differing power sources, each one leading to a different product concept.

Embodiment design

This is the stage at which one or more concept designs are translated in layout drawings, schematic drawings, mock-ups, visual models or prototypes for technical development and testing and for detailed evaluation by management [and] ... for eliciting feedback from potential customers or users. They can provide useful indicators of market acceptability.

In our hedge trimmer example, the embodiment design stage would involve deciding the configuration of the basic components (motor, cutter, shield, handle, controls etc.), performing calculations and tests on cutting rates, capabilities, safety issues and material considerations etc.

Detail design

At this stage the design selected at the embodiment stage is developed and optimised in detail. The form and arrangement of components is defined, the materials are specified and the design is adapted to reduce anticipated manufacturing problems or costs. For those companies with sophisticated CAD tools they are able to modify the computer based, virtual models made in the embodiment stage. [...]

The product is given final technical and market testing and presented as a final production prototype. At this stage the models of the product specify every detail. [...]

For the hedge trimmer, detail design might involve analysis of strength in components or ensuring that these parts can be moulded ... or otherwise shaped without difficulty. ... Test assemblies might be made using pre-production parts to highlight potential problems on the assembly line.

Models for concept designing

The essay by Mark Evans in the *Modelling Workbook* shows some of the different models employed at the different phases of product designing. At the concept stage these may include 2D concept sketches and diagrams, and 3D *principle-proving models* to test functional feasibility. These models need only be detailed enough to test or communicate the basic idea, principle or form. Thus, a good description of a 2D *concept sketch* is: 'a collection of visual cues sufficient to suggest the design to an informed observer'.

As the design moves to the embodiment and detail design stages, with the production of layout drawings, renderings, prototypes and engineering drawings the models become increasingly detailed and realistic. Increasingly, as the *Modelling Workbook* shows, these models are produced on computer-aided design and rapid prototyping systems.

W 1

> Either now or at the end of this section, spend 10 minutes referring back to Section 1 of the *Modelling Workbook* to remind yourself of the models – of a lawnmower, children's cutlery, and a grass trimmer – used at different phases of designing. Figures 1, 2 and 5 of the workbook show models used at the concept design phase.

Types of concept design

The examples in Block 1 and the *Modelling Workbook* focused on chairs and product design. Figures 6 to 12 illustrate some examples of the 2D and 3D models used in fashion, website and engineering design, as well as in furniture and product design, to generate and explore concept designs. You may notice the different factors addressed in the concepts from different design professions.

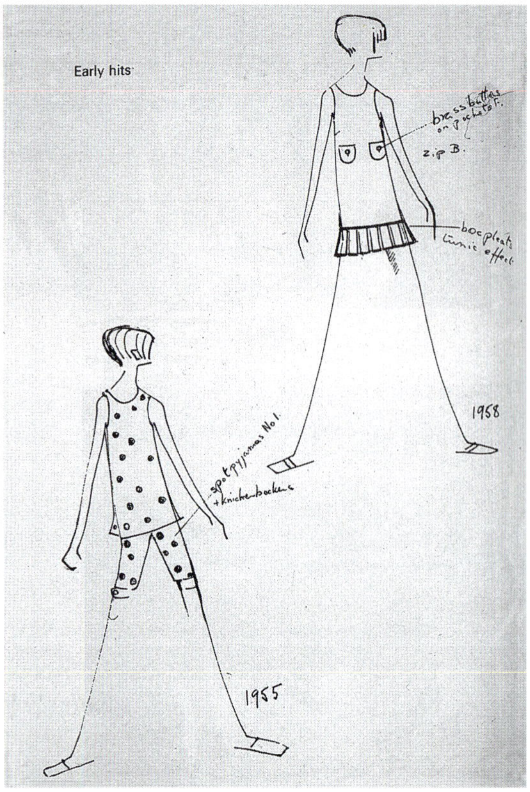

Figure 6 Sketches from the 1950s by the famous fashion designer Mary Quant focusing on visual design (how the garments look)

Fashion and some graphic (e.g. poster) designers are mainly concerned with *aesthetics* (visual appearance). Engineering designers are primarily concerned with *function* (how things work) and *manufacture* (how they are made). Other graphic designers (e.g. magazine, website) and product and industrial designers attempt to integrate *function*, *ergonomics* (usability) and *aesthetics*.

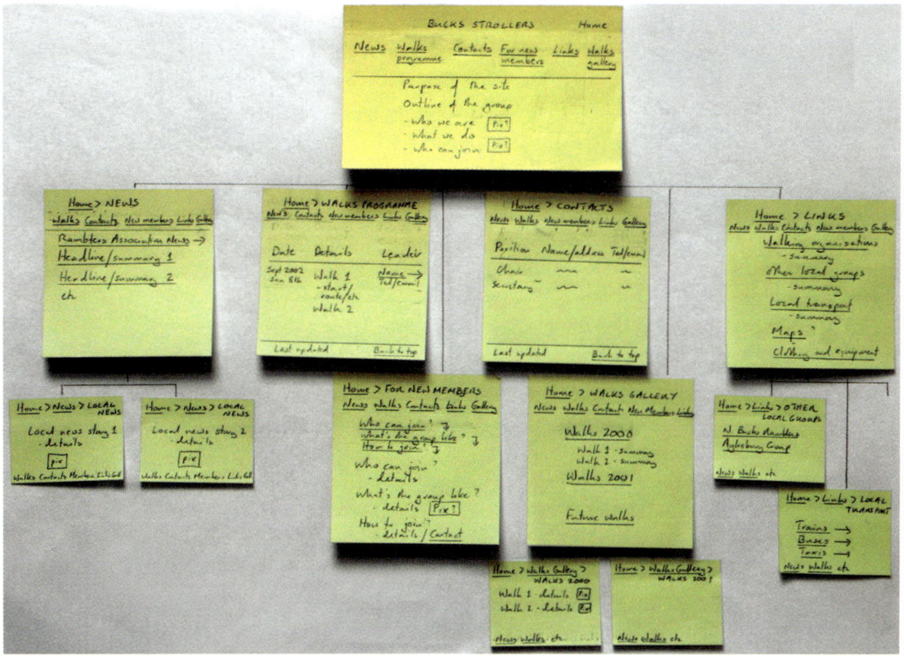

Figure 7 **Paper mock-up, using Post-It notes, of a website design for a local walkers' group, showing basic information structure and outline content**

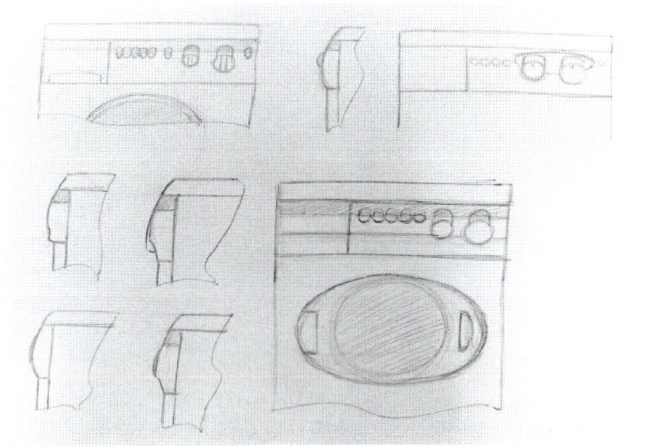

(a) Rough sketch

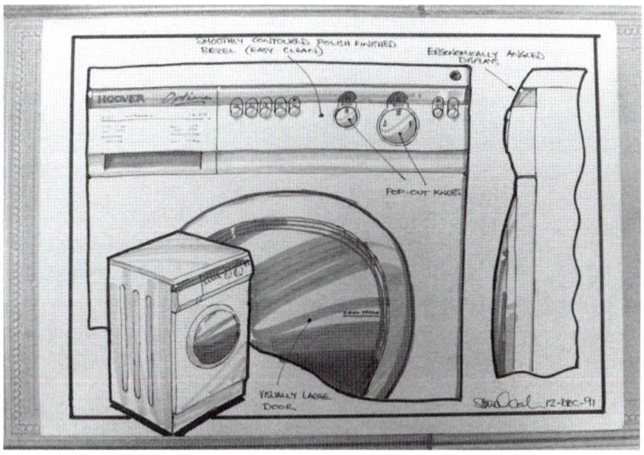

(b) perspective sketch

Figure 8 **Concept sketches produced during design of the control panel of a washing machine exploring visual, ergonomic and functional design (i.e. how the panel might look and work)**

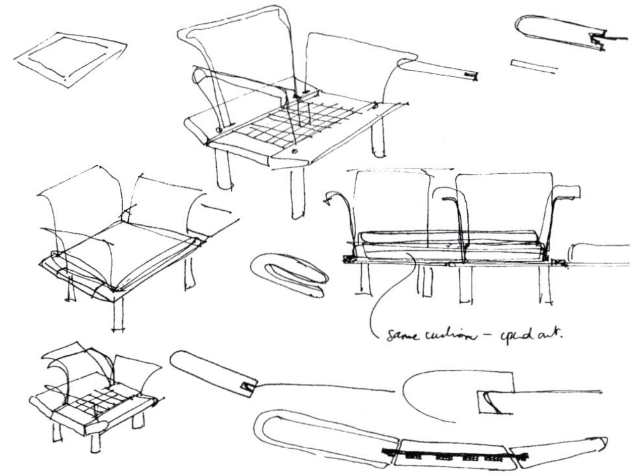

(a) Concept sketches

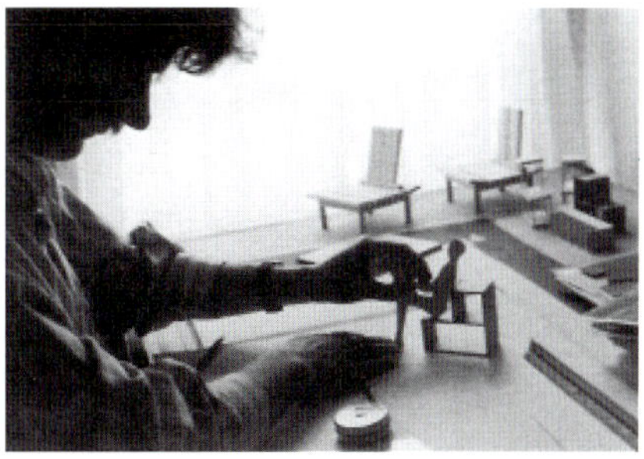

(b) scale models

Figure 9 **Functional and visual design of an adjustable sofa bed**

Note the exploration of elements of detail design at this early stage.

Figure 10 Computer model of a pickup concept vehicle for Ford

Although just a 'concept car' in the sense of exploring possible directions for vehicle design, this solid model provides a fairly realistic idea of how the vehicle might look and so has moved into embodiment design.

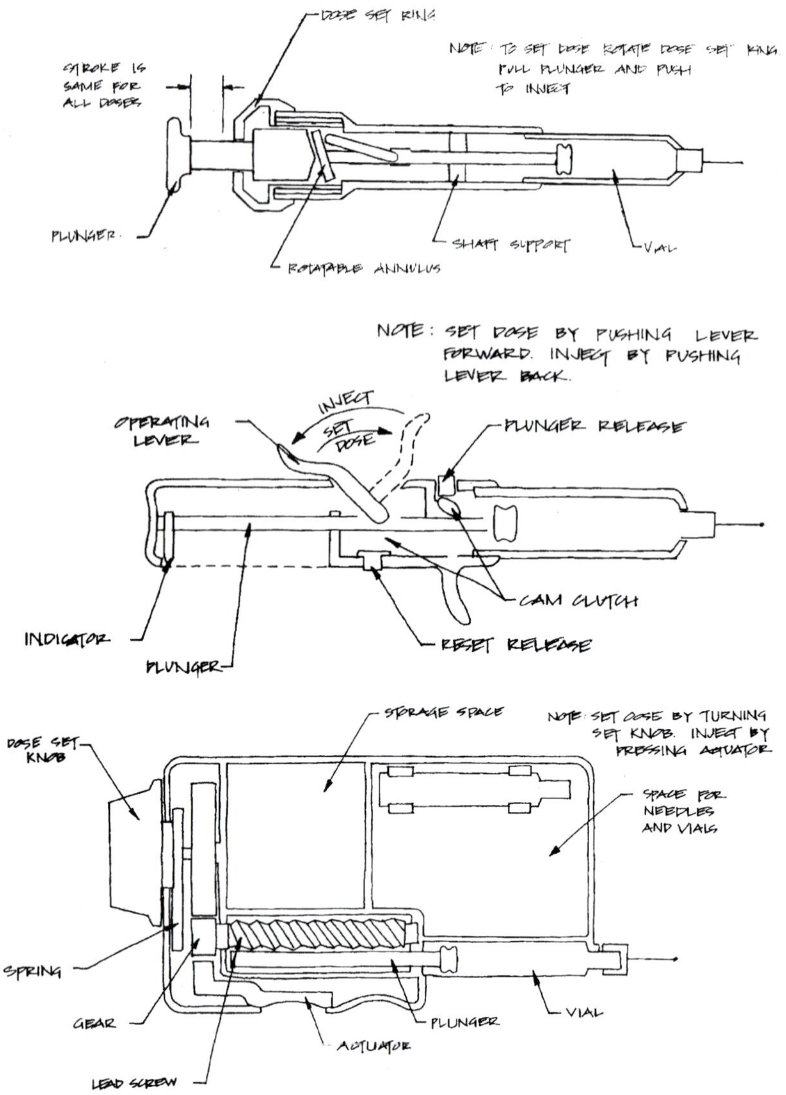

Concept E: Swash Ring

Concept F: Lever Set

Concept G: Dial Screw

Figure 11 Concept sketches of a syringe for outpatient use, focusing on functional design (i.e. how the device works)

a) First prize seven and under. The 'Robo-Caterpillar' 'to wake you up by crawling on you' (Steven Sutton, age 6 years)

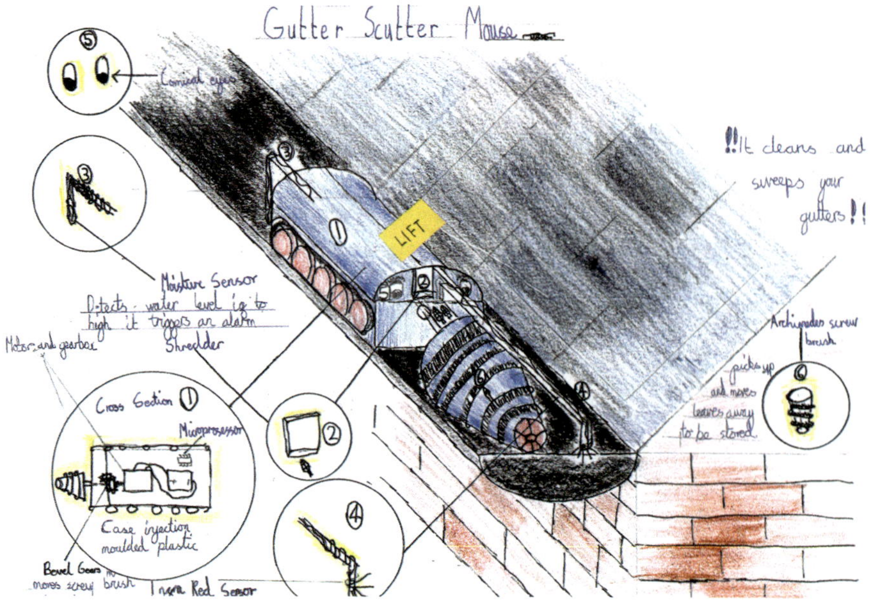

(b) The 'Gutter Scutter Mouse' robot to clear rain gutters (Stuart Gardiner, age 10 years)

Figure 12 Two entries to the *Blue Peter-Robofesta* children's competition to design 'a really useful robot'

What is involved in generating and modelling concepts may be very different in the different areas of design. A fashion designer or car stylist may be able to quickly produce some fairly realistic concept designs (e.g. Figures 6 and 10) based on their experience and creativity. In contrast, an engineering design team may have to do a lot of information gathering, research, mathematical and physical modelling before the concept or concepts emerge (e.g. Figures 2 and 11).

Exercise 3 Types of concept design

Why do you think there are differences between design professions in the time and effort involved in creating concept designs?

Spend a few minutes thinking about your answer before looking at my response below.

My response

The main reason is the degree of technical novelty and complexity that can be involved in the concepts from different design professions.

configuration
The arrangement of the parts of a design and how they are connected.

In areas such as fashion, textiles and graphics, and usually in furniture and ceramics, concept designs are usually different configurations of components, materials, information, colours and patterns based on existing technical principles (see e.g. Figures 4 and 6–9). The products concerned are usually relatively small and technologically uncomplicated. Individuals or small teams can often produce such concept designs quite rapidly. Even so, new technologies and materials are increasingly affecting the designs from these professions.

technical principles
The scientific or technological basis on which a device or system operates e.g. the physics of heat flow; mechanics of structures.

In engineering, and sometimes product design and architecture, there are often alternative technical principles to provide the solution to a design problem, (e.g. Moulton's fluid vehicle suspension, shown in Figure 2; Dyson's application of cyclone technology to cleaners, shown in Figure 5). Concept designing may involve exploring alternative technologies, or inventing and testing new technical principles, as well as producing alternative configurations. In the hedge trimmer example in Box 1, concept design involved exploring different principles for cutting hedges, such as rotating cutters, high pressure water jets, freezing or burning. The concept design of products, whether very small (micro-electronics) or very large (buildings, aircraft), may also be highly complex technologically. Such concept designing is thus an activity often requiring large teams and major research and modelling effort.

Levels of concept design

total product/system
(e.g .wave power device)

sub-system
(e.g. mooring cable)

component
(e.g. swivel joint)

What is involved in concept design also depends very much on the starting point. Concept design may start at the level of a total product or system, such as a motor vehicle, or at the level of a sub-system, such as the vehicle's suspension or its water pump, or at the level of a component such as a pump bearing.

Take for example an engineering problem given at the total product/system level: the concept design of devices to harness the energy of the waves. There are very many ways in which wave-power might be harnessed, and so this problem offers much scope for invention and creative design ideas. Indeed, there are over a thousand proposals for utilising wave energy in the patent literature.

One of the simplest wave-power concepts is an oscillating water column that uses the rise and fall of the waves to drive air through a turbine, and many variants of this idea are possible. Two variants of the oscillating water column, and three other concept designs for wave-power devices, are shown in Figure 13.

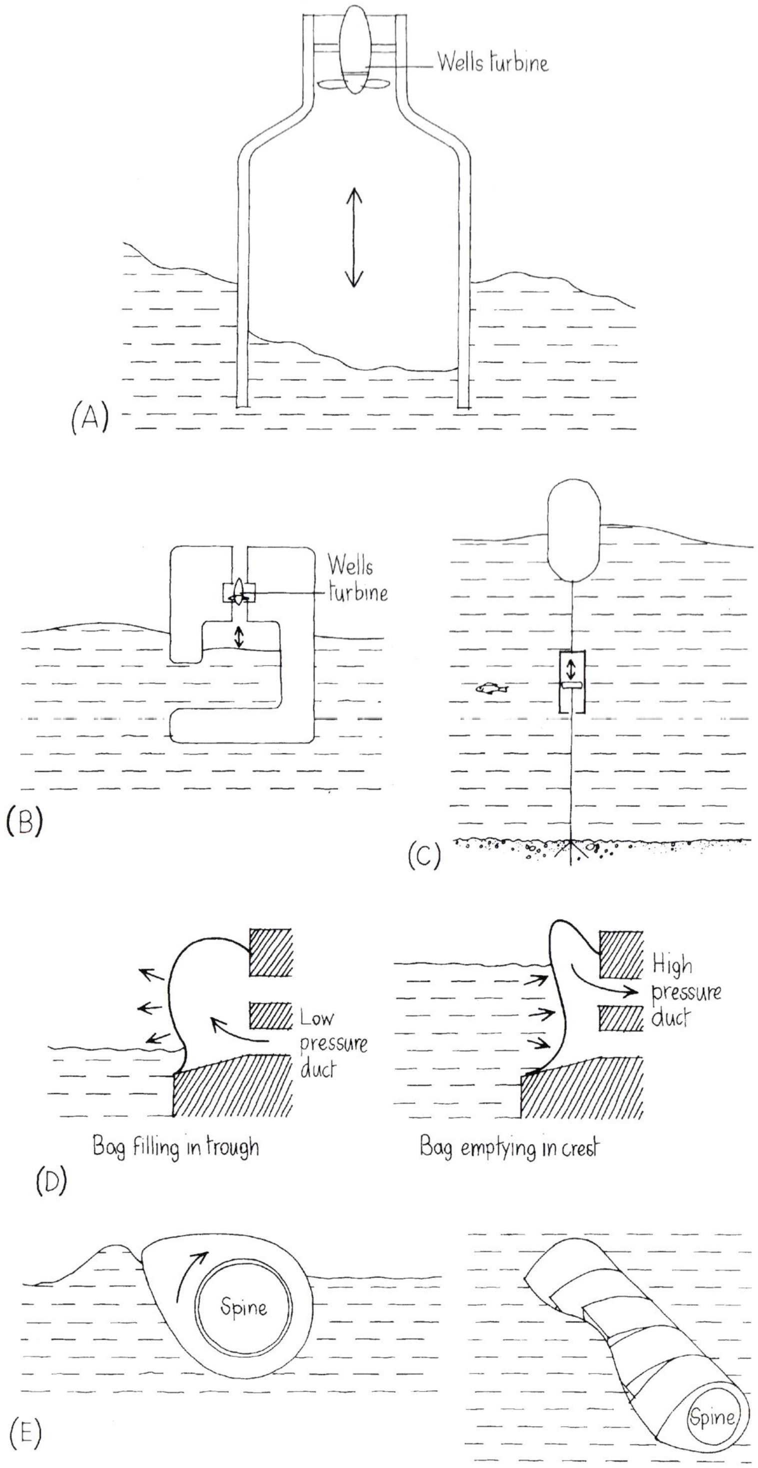

(a) Simple oscillating water column

Water rising and falling in the space inside the device pumps air through a (Wells) turbine, which accepts fluid flow in either direction.

(b) National Engineering Laboratory's oscillating water column wave energy converter

(c) Bobbing buoy

Waves passing over the buoy increase and reduce the force in the mooring cable thus operating a hydraulic pump that in turn drives a turbine.

d) Lancaster flexible bag

Wave crests squash the air bag, which acts like bellows forcing air through a turbine: the bag refills in wave troughs.

(e) Salter's 'Duck'

Cam-like bodies oscillate about a central spine as the waves pass thus operating hydraulic pumps inside the device, which in turn drive turbines.

Figure 13 Concept designs for a total product/system – wave-power devices

Note that the two oscillating water column concept designs, Figures 13(a) and (b), share the same basic *technical principle* of operation – air being forced through a two-way turbine by the motion of the waves. They are variants of the same concept design. The Lancaster bag, Figure 13(d), operates on a similar technical principle. But it has different *working principles*, having a bag to contain the air and being fixed to the seashore rather than floating. It is thus a different *concept design*. The bobbing buoy, Figure 13(c), operates on yet another technical principle – a floating body fixed to the seabed whose motion in the waves drives a hydraulic pump. It is the many combinations of possible technical and working principles that accounts for the large number of wave power concept designs.

Designers are not often given a problem at the total product/system level. As you've seen, most concept design isn't concerned with creating *new* technical principles, but with providing alternative configurations for an *existing* principle. A typical engineering problem would be to design a component for one of the wave-power devices to meet a given specification. Figure 14 shows three designs for swivel joints that might be used for connecting undersea cables on a wave-power device. The differences between these designs are not obvious unless you have the engineering knowledge to interpret the drawings. For example, Design 1 with a plain bearing is less complex and easier to manufacture than Designs 2 and 3 that have internal ball bearings giving them smoother operation. From the viewpoint of the engineers who produced them, designing these joints required a considerable amount of creative work at the concept stage involving 'much freehand sketching' before they could be drawn in detail, as shown here, using a computer-aided drafting system (Shahin, 1988).

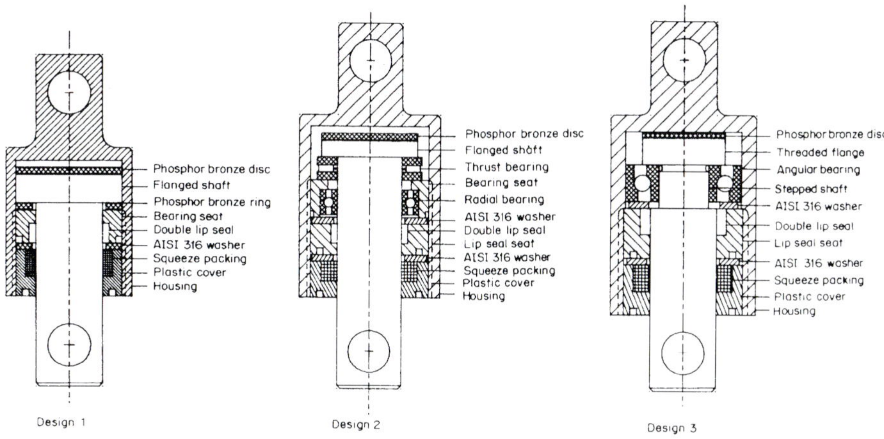

Figure 14 Alternative designs for a swivel joint

The main differences are in the choice and configuration of the bearings and seals.

The wave power devices and the swivel joints are extreme examples. Most concept designing is located somewhere between inventing whole new systems and designing minor components. Indeed, in this block I am mainly concerned with creative thinking applied to the design of new or improved versions of relatively simple total products, such as bicycles or vacuum cleaners, or major sub-systems or components of such products. The point of giving these extreme examples is to show that there are alternative solutions to a design problem at every level from the most general to the most detailed. The creative thinking needed to explore the solution space and to produce worthwhile concept designs is needed at all stages and all levels of design.

solution space
A notional space within which lie all the possible solutions to a design problem – the set of all possible solutions.

1.3.3 Ideas, concepts and inventions

Turning back to the examples I've discussed so far they indicate some important distinctions.

idea
The fundamental thought or notion for solving a problem.

The basis of a concept design is an idea in the mind of the designer or team. Examples of such ideas that you've seen include: a flexible bag wave power device; an air-filled tyre; a bike that folds into a stick with wheels at one end; and a squid shaped, spaceship-like lemon squeezer. An idea may involve an existing or a new technical principle for the particular function concerned. Such ideas may emerge during the task or problem clarification phase of designing, showing that creativity is involved at this initial phase too.

concept design
The physical embodiment of an idea in a sketch or model.

The idea then has to take shape, to be *embodied* in specific physical configuration as one or more concept designs. A mock-up of a pneumatic tyre made from wood, cloth and rubber tubing; sketches of possible structures for a bike that could fold into a stick shape; and sketches of concepts for a lemon squeezer with a squid shaped head are all examples of this. As one designer said, 'a successful design is one that perfectly encapsulates the idea that inspired it. But each designer will interpret the idea in a different way'. The chosen concept is then developed through the stages of embodiment and detailed design until suitable for production and practical use.

invention
An idea or concept for a new or improved device, product or process; something that previously did not exist .

patent
Legal protection for an invention that is proved to be new, is not obvious, and is capable of being applied as a practical product or process.

If the idea or concept design is based on *a novel technical principle for its particular function or manufacture* (even if that principle has been applied in other devices or systems), it may count as an invention. Invention is, however, an imprecise term and only has clear-cut legal meaning if the invention is patented. For an invention to qualify for a patent – giving the inventor a monopoly right for a limited period to exploit the invention – it has to fulfil four conditions:

1. It must be *new* – the invention must never before have been made public anywhere in the world.
2. It must involve a so-called *'inventive step'* – when compared to what is already known, the invention should not be considered as 'obvious' to someone knowledgeable in the subject.
3. It must be capable of *industrial application* – the invention must take the form of a practical device, a new material or an industrial process. It should be described in sufficient detail (e.g. in words and drawings) for a third party to be able to reproduce it.
4. It must not be *'excluded'* – e.g. scientific discoveries, computer programs and artistic works cannot be patented under UK law.

Thus the 'Hydrolastic' suspension system or pneumatic cycle tyres, after they had been demonstrated as working prototypes, were plainly patentable inventions. But, because of the requirement for newness, Dunlop was only granted a patent for his invention after a dispute over Thompson's earlier invention of pneumatic tyres for heavy vehicles had been resolved.

On the other hand, Galliano's collections of fashion clothing do not count as inventions nor could they be patented, as they do not involve any form of non-obvious technical advance. Their novelty lies in the new visual configurations of materials and the textures, colours and patterns that he and his team create for each season. (The designs are covered by a different form of legal protection – copyright.)

Only a minority of inventions result in technical innovations. Only a few inventions, even patented ones, undergo development to the production stage. And, even then, not all production designs are introduced onto the market or otherwise brought into use. Only at the point of *initial commercial introduction* (or adoption into use) is an invention said to have become a technical innovation. I shall discuss the factors that influence the process of technical innovation in Section 6.

innovation
The first introduction or use of something new.

technical innovation
The commercial introduction of an invention.

Exercise 4 Ideas, concepts, invention and technical innovation

Reread the brief descriptions of the creation of (a) Sanders' folding bicycle and (b) Starck's Juicy Salif lemon squeezer in Section 1.3.1. For each design identify:

- the basic (underlying) idea;
- their technical principle (of operation);
- concept design(s) that were produced;
- whether the designs were patentable inventions.
- if the final products were technical innovations?

Spend 5–10 minutes thinking about your answer and making notes in your workfile before looking at my response below.

My response

(a) Sanders' basic *idea* was a bike that folded like a baby buggy into a stick-like form.

The *technical principle* was some form of mechanism applied to a bicycle that could provide the folded form.

Several *concept designs* embodying different folding mechanisms and configurations were sketched and drawn.

Although the idea of a folding bicycle is not new, Sanders' idea of a bike that can fold into a stick-like form was novel and not obvious. He developed his idea into a working prototype without disclosing it to anyone else and so could *patent* his *invention*. The bicycle was marketed commercially and so can be considered to be a *technical innovation*.

(b) Starck's basic *idea* was a novel form of a lemon squeezer for a manufacturer of household products that sells to design-conscious consumers.

The idea involves the same essential *technical principle* as a conventional lemon squeezer.

The idea evolved into a *concept design* with a configuration rather like a space ship.

Starck's Juicy Salif is highly original in its form and appearance, but does not offer a new way of squeezing lemons (or of making lemon squeezers) so can't be patented.

Although launched onto the market, because it uses existing principles the design can't be considered as a *technical innovation* (although it might be described as a product innovation).

From the above discussion I concluded that Starck's Juicy Salif wasn't a patentable invention. But as the design has a novel shape it would automatically be protected from copying in the UK by Design Right, a special form of copyright that applies to 3D products. It could also be better protected under the law on Design Registration, which gives the designer the exclusive right to use the design for up to 25 years within a country or group of countries

Design Registration
Legal protection for a product or component that appeals because of its *visual appearance*. It must be new in configuration, shape, and/or surface pattern and be perceived to have an 'individual character'.

1.4 Design thinking

So far I've focused on creativity in design. But just being creative will not normally lead to satisfactory concept designs. Block 1 stressed that any new concepts should lead to designs that work, are capable of being made, don't cost too much and satisfy their producers and consumers. Let's consider the other types of thinking required in design and how you might apply them during the designing process.

1.4.1 Exploration, generation and selection

You've seen that creative thinking is required throughout the design process and that creativity is especially important in the concept phase of designing because that is where the basic idea or solution concept is conceived.

But, as my earlier examples, such as Moulton's Hydrolastic suspension, hinted, in many areas of design more time and effort may be required to explore the problem in question (perhaps by gathering information or doing calculations) than in creating solutions. As you will see, this problem exploration sometimes also provides the basic solution idea. Then, as much effort may be required to decide between possible alternative solutions as was needed to generate them in the first place.

This gives us a clue to three essential, but very different, types of thinking required in design (and indeed any kind of problem solving).

In Block 1, Section 9.1 these three types were described as:

- design specification;
- design generation; and
- design evaluation.

More generally they have been described as:

- problem exploration;
- idea generation; and
- solution selection.

Another way in which to look at these different kinds of thinking is in terms of their effect within the designing process. Just as creative thinking to generate ideas is usually involved at all stages of design from concept to detail, so are the other types of thinking. In fact the whole designing process may be viewed as successive and repeated cycles through the three types of thinking, as shown in Figure 15.

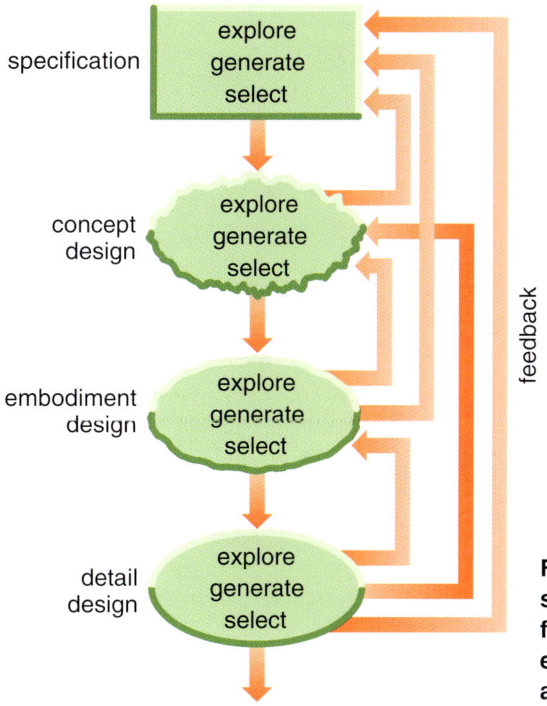

Figure 15 Designing process showing repeated cycles, with feedback loops, of problem exploration, idea generation and solution selection

SAQ 1

Look back at Figures 86 and 87 in Block 1.

Are there any important differences between the design cycle models of designing in Block 1 and Figure 15 presented here?

My response to this and other self assessment questions (SAQs) can be found at the end of this block.

Let us look at an example (adapted from Pugh, 1991) of the three kinds of thinking (i.e. exploration/specification; generation; evaluation/selection) in the concept phase of an engineering design project.

Case study Concept design of a car horn

The brief is to design a new type of car horn. First the task given in the brief has to be analysed and defined more precisely. The design team attempts to identify the performance requirements for the new design, the range of conditions to which it will be exposed, any constraints for size, weight, cost and so on, to be incorporated into the *product design specification* (see Block 2, Section 10). The design team also gathers information on existing designs of car horn and other audible warning devices, and identifies different technical principles for producing sound. This problem *exploration* (also termed 'preparation' or 'immersion' in the problem) provides the context for generating solutions.

The next step is the most creative part, in which the different ideas for producing sound are further explored and embodied in alternative concept designs. Figure 16 shows the conventional car horn (concept 1, top left) plus thirteen alternative concepts produced by a group of trainee designers. I shall leave until later discussion about how designers actually generate ideas and concepts such as these.

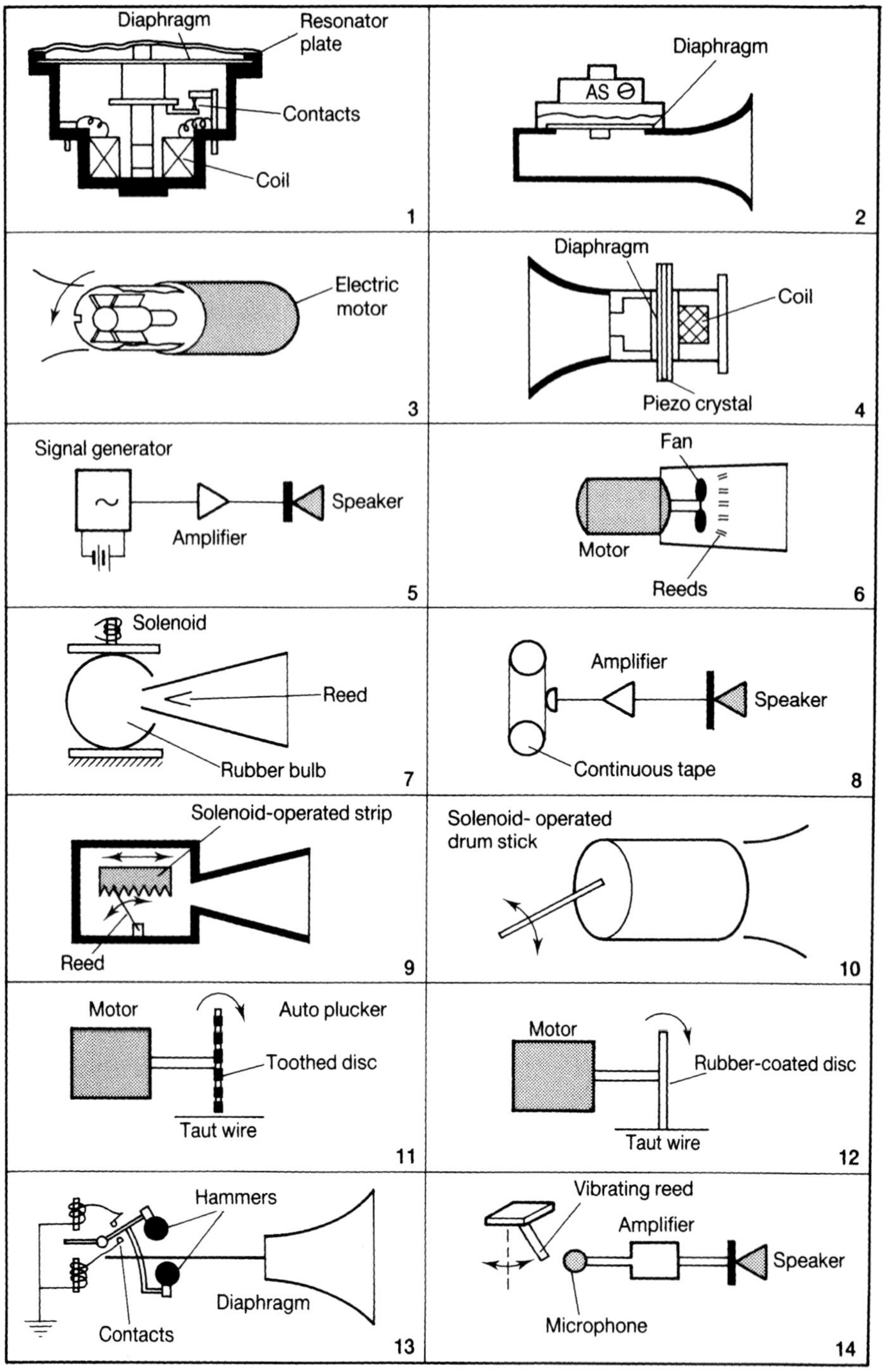

Figure 16 Alternative concept designs for a new type of car horn

The conventional car horn is concept 1. Different technical principles are employed to produce sound in the different concepts e.g. concepts 5, 8 and 14 are electronic devices with different sound sources (Pugh, 1991, p. 82).

Finally it's necessary to evaluate the different concepts in order to *select* one (or more) for further development. One technique for doing this (that we shall return to in Section 5.3.3 and in an exercise on the DVD) is to compare each concept with the existing design using criteria from the specification. Figure 17 is an evaluation or selection matrix showing the specification criteria against which to evaluate the various car horn concepts. It reveals concept 5 (an electronic device) as having many advantages and few drawbacks compared to the existing design (concept 1), and so seems like a promising concept for designing a new type of car horn. In this case the evaluation was done by means of group judgement, but as you will see later, choices between concepts often cannot be made without considerable mathematical analysis and testing of mock-ups and prototypes.

Criteria \ Concept	1	2	3	4	5	6	7	8	9	10	11	12	13	14
Ease of achieving 105–125 dBA	**D**	S	-	NOT EVALUATED	+	–	+	+	-	-	-	-	S	+
Ease of achieving 2000–5000 Hz	**A**	S	S		+	S	S	+	S	-	-	-	S	+
Resistance to corrosion, erosion and water	**T**	-	-		S	-	-	S	-	+	-	-	-	S
Resistance to vibration, shock and acceleration	**U**	S	-		S	-	S	-	-	S	-	-	-	-
Resistance to temperature	**M**	S	-		S	-	-	-	S	S	-	-	S	S
Response time		S	-		+	-	-	-	-	S	-	-	-	-
Complexity: number of stages		-	+		S	+	+	-	-	-	+	+	-	-
Power consumption		-	-		+	-	-	+	-	-	-	-	S	+
Ease of maintenance		S	+		+	+	+	-	-	S	+	+	S	-
Weight		-	-		+	-	-	-	S	-	-	-	-	+
Size		-	-		S	-	-	-	-	-	-	-	-	-
Number of parts		S	S		+	S	S	-	-	+	-	-	S	-
Life in service		S	-		+	-	S	-	-	-	-	-	-	-
Manufacturing cost		-	S		-	+	+	-	-	S	-	-	-	-
Ease of installation		S	S		S	S	+	-	S	-	-	-	S	-
Shelf life		S	S		S	S	-	-	S	S	S	S	S	S
Total: +	0	0	2		8	3	5	3	0	2	2	2	0	4
Total: -	0	6	9		1	9	7	12	11	8	13	13	8	9
Total + less Total -	**0**	**-6**	**-7**		**7**	**-6**	**-2**	**-9**	**-11**	**-6**	**-11**	**-11**	**-8**	**-5**

Key: + means better than the conventional design (the DATUM, column 1)

\- means worse than the conventional design

S means the same as the conventional design or cannot assess

Figure 17 Evaluation (or selection) matrix for comparing new car horn concepts with the conventional design (adapted from Pugh, 1991, p. 83)

1.4.2 Heuristic and systematic designing

systematic design
A step-by-step design process that encourages designers to methodically explore the problem and the solution space to find an optimum design.

heuristic design
Designing using approaches or 'rules of thumb' (e.g. iteration) that guide problem-solving, but do not guarantee an optimum design.

The car horn example – and the process represented in Figures 16 and 17 – are based on a systematic design approach. A systematic search of the *solution space* (see Section 1.3.2) is especially worthwhile in engineering problems such as this, because concept designs can involve alternative technical principles as well as different configurations. Nevertheless, most experienced designers don't actually work in such a formal, systematic way. Experienced designers typically work from problem to solution in a way that appears 'intuitive' without overtly generating and evaluating alternative concepts. This is because their knowledge and experience will provide them with approaches and 'rules of thumb' that enable them to home in on promising concepts in the solution space. This is known as a heuristic design approach to problem solving. A common heuristic approach is to break a complex problem into successively smaller problems until you are left with a set of problems, which can be solved. Another, powerful, heuristic approach is that of 'successive approximation' or *iteration*, that is producing a rough draft solution and gradually improving it. Indeed, as you will see later in this block, many experienced designers start with an idea for a solution and use it to explore the problem in hand, leading to new insights or modified ideas. This is the so-called 'solution-focused' approach, illustrated by the design case studies that we will explore in Sections 2 and 3.

Those who favour a systematic approach argue that creating a design by heuristic methods has drawbacks, as better alternatives within the solution space may be missed. But whatever the approach, the overall effect of the process is to narrow down from an ill-defined problem with many possible solutions to a single solution specified in enough detail for manufacture.

1.4.3 Variant and innovative designing

The best approach to take depends on the design problem. Many design problems are sufficiently well understood to be solved mainly from past experience and need relatively little creativity. Most designing produces *variants* of existing products (e.g. a redesigned web page, an improved car gearbox, or a new model of washing machine). However, in this block I'm concerned mainly with design problems whose solution offers the scope for high levels of creative thinking leading to *innovative* concept designs and products, like Moulton's Hydrolastic suspension system or the Cyclone cleaner.

Exercise 5 Alarm clock concepts

Suppose you were asked to produce alternative concept designs for an alarm clock to be marketed in five years time.

(a) Consider how the context in which the clock will be developed and used might have changed in five years from now. What *technical* changes are taking place in timing/alarm devices and systems? What *social* and *human* trends (e.g. in work patterns) might affect the demand for and use of alarm clocks?

(b) Generate some ideas for the alarm clock taking into account the *source of power* used to drive the clock, the way in which the *time is indicated* and the *type of alarm* system. These ideas may be described in words.

(c) Embody your ideas in at least two alternative concept designs – with an annotated diagram or a rough sketch.

Spend about 15 minutes noting down ideas and sketching your concepts in your workfile before looking at my response below. If possible, use the sketching skills you've learned from the *Modelling Workbook* Sections 4 and 5. However, your standard of sketching need be no better than the Robo-Caterpillar or the Gutter Scutter robots in Figure 12.

Your attempts at this and other exercises will be useful when doing the TMA for this block; take a look at the Block 3 TMA if you have not done so already. Use your workfile to keep any notes and sketches you make.

My response

(a) Relevant *technical* changes include miniaturisation of products; 'intelligent' products; synthesised sounds; radio- and internet-controlled clocks, and alarm devices/computers combined with many other products e.g. mobile phones, microwave ovens. These developments offer the potential for low cost, interactive, electronic devices.

Relevant *social trends* include '24/7' work patterns; more single person households; irregular work/leisure patterns; etc. These suggest a greater need for alarm devices at home, work and elsewhere.

(b) Different possible *sources of power* include: wound spring, falling weight, pendulum, electric battery, solar cells, wind generator.

Different ways of indicating the *time* include: hands, electronic display, audio output, projection display, visual changes, etc.

Different types of *alarm* include: buzzer, bell, various sounds (e.g. music, speech, cock crowing), vibrations, flashing light, scents, etc.

Combinations of these different options give a large number of possible product ideas, from a conventional spring-driven alarm clock to an electronic clock that speaks the time and tells you when it is time to get up.

My chosen ideas, derived partly by thinking of the technical and social trends and partly from the technical combinations, are:

a *mobile phone alarm* that you programme like an alarm call telephone service. There is a choice of ringtones, or a synthesised voice. You have to say you are awake, or tell it to call later, before the alarm stops;

an *internet connected alarm* clock connected to the Internet for weather reports, music etc;

a *toaster alarm* that wakes you up by the sound and smell of toast being made;

a *soft alarm* clock that has a soft case which is thrown against the wall to stop the alarm.

(c) Some concept designs, by a colleague, embodying these ideas are shown in Figure 18 overleaf.

The method of generating many ideas through combinations of different ways of achieving the functions of an object that I used in answering part (b) of the exercise is called *morphological analysis*. It is one of the creative problem solving techniques in Section 5.3 of this block and on the DVD.

SAQ 2

Define each of the terms in the following three groups. Then explain the relationships between the terms within each group:

(a) idea, concept design, technical principle, configuration, creativity;

(b) invention, patent, technical innovation;

(c) solution space, systematic design, heuristic design.

Before you read on, look back over this first section and in your workfile make a summary of its key points in about five or six bullet points. Do this before looking at my own summary overleaf, then compare your key points with mine. You should try and do this at the end of each section as you work through the block.

(a) Toaster alarm

(b) Mobile phone alarm

(c) Internet connected alarm

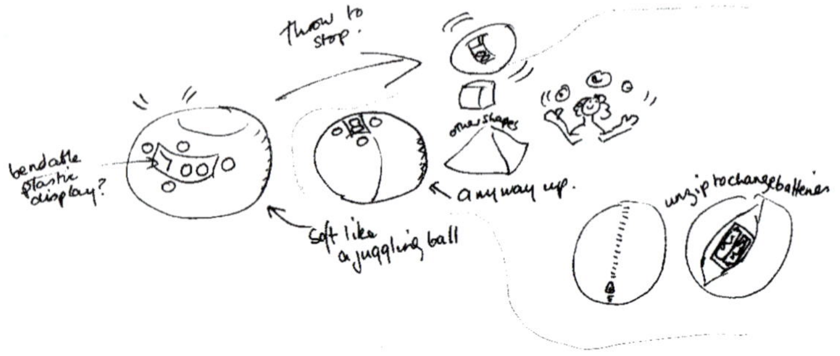

(d) Soft alarm

Figure 18 Future alarm clock concept designs by Georgy Holden

Key points of Section 1

Key terms used throughout the block are defined, including:

> creativity; idea; technical principle; configuration; concept design; invention, patent, technical innovation, solution space; systematic design, heuristic design.

Creativity is required *throughout* the designing process, but especially during concept design.

Concept designs are based upon an *idea* involving a particular *technical principle* of operation and embody the idea in a specific arrangement or *configuration*. If the concept is based on a novel principle for its particular function or manufacture, it may count as a *patentable invention*.

In some design professions (e.g. engineering) concept design may involve exploring alternative technologies or inventing and testing new technical principles, as well as designing alternative configurations. In other design professions (e.g. ceramics) concept design involves creating new configurations based on existing technical principles and materials.

Creative concept designing typically involves *new ideas* inspired by other objects; 2D and/or 3D *modelling*; sustained *effort*; relevant *knowledge*; an understanding of *market*, *materials* and *manufacturing* constraints; and often *teamwork*.

The designing process may be viewed as successive and repeated cycles through three types of thinking – *exploration/specification*; *generation*; and *evaluation/selection*. This process may be undertaken using *heuristic* and/or *systematic* approaches.

(Learning outcomes 1.1–1.4 and 2.2)

2 The creative process

So far I have discussed creativity in design with only some hints on how novel ideas and concepts are produced – for example, that ideas are often inspired by objects from a different or related field. In Sections 2 to 4 I shall be using several examples and case studies – mainly from the development of bicycles – to draw out some of the principles and theories of creative thinking in design and invention.

Here in Section 2, I focus on the creative process in general.

2.1 Bicycles – historical examples of creativity

I'll start by looking at two landmark developments in the history of bicycle design.

The modern bicycle evolved from a series of inventions and designs that originated in the early 1800s. By about 1890 the bicycle had developed into its classic diamond frame, rear chain driven form.

Case study The 'Rover' bicycle

A key step in the evolution of the classic diamond frame design was the 'Rover' bicycle designed by John Kemp Starley in 1884.

It's important to understand that the Rover did not emerge in one giant leap 'out of the blue', but was the result of nearly twenty years of cycle design evolution and component innovation. These developments included wire-spoked wheels, ball bearings, hollow frames, and a huge variety of designs with different frame, wheel, transmission and steering configurations. These designs included the fast but dangerous 'Ordinary' bicycles ('penny farthings', see Figure 19(a)) and a variety of 'safety' cycles that prevented the rider being thrown over the handlebars if the front wheel hit an obstacle. Safety bicycles and tricycles at the time included front and rear chain driven designs (e.g. Figure 19(b)). In conceiving the Rover bicycle, like most creative designers, Starley put together in a new and better way what had gone before.

What was the thinking that led Starley to the basic idea underlying the design of the Rover? A lecture that Starley gave to the Society of Arts in London in 1898 gives us some clues.

> I felt the time had arrived for solving the problem of the cycle […]. I therefore turned my attention solely to the perfection and manufacture of the Rover bicycle.
>
> The main principles which guided me in making this machine were to place the rider at the proper distance from the ground; to connect the cranks with the driving wheel in such a way that the gearing could be varied as desired; to place the seat in the right position in relation to the pedals, and constructed so that the saddle could be either laterally or vertically adjusted at will; to place the handles in such a position in relation to the seat that the rider could exert the greatest force upon the pedals with the least amount of fatigue; and to make them adjustable also.
>
> (Starley, 1898, p. 608)

(a) Typical 'Ordinary' or 'penny farthing' bicycle of the 1870s

Front wheel sizes ranged from 40 to 60 inches (1 to 1.5 metres).

(b) Lawson's 'Bicyclette' of 1879

This 'safety' design included rear chain drive but retained the large front wheel of the 'Ordinary' bicycle.

Figure 19 Bicycle designs of the 1870s

We see here some of the essential features of creative design thinking:

1 dissatisfaction with what already exists – what in Block 2 was called ‘constructive discontent’.

2 a problem frame – a particular view or understanding of the problem on which to base an improved design. In Starley’s case he saw the problem as how to enable the rider to exert the maximum downward force on the pedals with least fatigue by best positioning of the saddle, handlebars and pedals.

problem frame
A problem-solver’s (e.g. a designer’s) particular viewpoint on or understanding of a problem arising from their knowledge and experience and/or analysis and exploration of it. (i.e. ‘I see the problem as. …’)

Starley went on in his lecture to explain how he tackled the problem:

> The illustration which I now give [Figure 20] will show what was running in my mind at this time. I had been considering what a man pedalling a bicycle could be compared to … it largely resembled walking up a ladder, but … whereas the pedals went down in the former, the man went up in the latter. I therefore had to determine where the handles should be placed to enable him to bring the whole of his weight on to the pedals. … It was … the handle-bar which compelled me to adopt the present form of machine, as I could not get it sufficiently forward by the other type. It will be seen by the position of the handle-bar on the ordinary [penny-farthing] bicycle, that it was utterly useless and imperfect for this purpose [Figures 19(a) and 20].
>
> (Starley, 1898, p. 610)

Figure 20 Starley’s ladder analogy

Starley used this analogy to establish the configuration of the Rover’s handlebar, saddle, and pedals to permit best use of the rider’s muscle power. On an Ordinary (penny farthing) bicycle (shown in outline) the rider is forced to sit more upright to grip the handlebars and so cannot exert maximum force on the pedals. (Adapted from Starley, 1898, p. 610).

You will notice that Starley produced his ergonomically improved design by means of an analogy – someone climbing a ladder.

analogy
Using an idea, object or situation that is similar in some way to what is under consideration.

The first prototype of the Rover was built in 1884. It had indirect steering and a 36-inch (90 cm) front wheel (Figure 21). In his first design Starley seemed unable to completely break away from the ‘penny farthing’ form. It was not until the third prototype of 1885 (Figure 22) that a diamond-shaped frame and direct steering to sloping, and thus self-centring, front forks appeared. These were ideas first featured in Humber’s tricycle and safety bicycle of 1884. Like many creative individuals, Starley employed adaptation of existing solutions to develop his own designs.

adaptation
Modification of an existing product or component to produce a new design.

Figure 21 First prototype of the Rover bicycle with a braced single-backbone frame, built in 1884

Figure 22 Third prototype of the Rover bicycle with a diamond frame built in 1885

By 1885, therefore, the basic configuration and most of the components of the modern bicycle had been established. It still remained to develop a triangulated frame by adopting straight tubes and adding a seat tube (I'll discuss the benefits of triangulated structures in the next section). By the mid-1890s most cycles were of this standard design (Figure 23) with only innovations such as the use of lightweight materials, and components such as gears and electric lamps, to distinguish it from contemporary bicycle designs.

Figure 23 Fully evolved Rover bicycle of the mid 1890s

Case study The pneumatic tyre

The pneumatic tyre was probably the most significant of the technical innovations to be applied to the diamond frame bicycle, displacing earlier solid rubber and cushion tyres. The pneumatic tyre not only increased the rider's comfort by absorbing road bumps, but also reduced effort by minimising the jarring of the unsprung mass of the cycle and rider.

How did the idea of the pneumatic tyre arise?

In Section 1.3, I quoted Dunlop's account of his invention of the pneumatic cycle tyre as an example of creativity:

'For a period of over twenty years Dunlop had considered … various forms of spring wheels'

> Dunlop's mind … had been considering the problem of vehicle vibration. For a period of over twenty years Dunlop had considered, and then abandoned as impractical, various forms of spring wheels, including the idea of a flexible steel band supported by rollers. Eventually he thought of the idea of a wheel with an air-filled tyre, which he envisaged as being made of cloth, rubber and wood.

Here we see that attempts to understand and solve a problem may be needed over a long period before the basic idea for its solution occurs. This is sometimes described as the *preparation* phase of the creative process, in which creators *immerse* themselves in a task, *explore* the problem and possible solutions.

But why an air-filled rubber tyre?

Although Dunlop does not say how he got the idea, it seems likely that it stemmed from his experience of making rubber appliances during his work as a veterinary surgeon. This would mean that his thinking had involved *adaptation* of existing objects (veterinary appliances), or more strictly transfer of ideas or objects from one field (veterinary appliances) to another (vehicle wheels). Another version is that he got the idea from seeing animal intestines. This would have meant Dunlop used an *analogy*, in this case a *biological analogy* (intestines looking like an inflated tyre), for his key idea.

transfer
Application of ideas, principles or designs taken from one field to a different field.

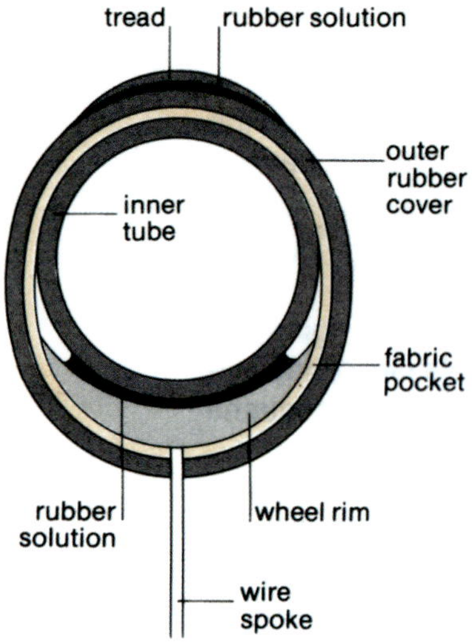

Figure 24 Dunlop's original pneumatic tyre of 1888

This design was later known as the 'mummy' tyre.

You've already read that Dunlop quickly tested his basic idea by conducting an experiment using a solid disc of wood to which he nailed a pocket of linen that enclosed an inflated rubber tube (presumably from his veterinary work). For new technical ideas, *experimentation* with a 'principle-proving model' – a mock-up that models the essential technical principle – or even a working prototype is often essential to test basic feasibility.

Encouragement for the practical development of the air tyre came from Dunlop's son who wanted to beat his friends in cycle races. Dunlop's experimentation progressed to fitting a pair of wheels with inflatable tyres to the boy's tricycle. For this key experiment Dunlop made wooden-rimmed wheels with the rubber tube contained in a canvas pocket wired and glued to the rim. The canvas was protected by an outer wrapping of sheet rubber stuck on with rubber solution. A simple non-return valve in the inner tube allowed inflation, but no deflation (Figure 24).

The prototype tricycle showed that pneumatic tyres were workable and so Dunlop proceeded to fit a second tricycle with his new tyres, which was ridden by a friend. Dunlop's friend was immediately struck by the smooth running and speed of the machine and encouraged Dunlop to exploit his idea. In 1888 Dunlop applied for a patent on his invention.

Thus, once Dunlop had conceived the basic idea of an air-filled tyre he performed experiments to test and develop it into a practical design. In the creative process he undertook what has been called *verification*.

2.2 Theories of creativity

These two historical examples from the field of bicycles exhibit many of the features of the creative process as it applies across different fields of human activity.

2.2.1 Process theories

One of the old, but still widely used, models of the creative process is that formulated by the psychologist Graham Wallas in *The Art of Thought* (1926) based on accounts by scientists and mathematicians of the stages they went through when trying to solve difficult problems. A similar model was proposed by a historian Abbott P. Usher, after examining the thinking behind mechanical inventions (Usher, 1954).

Wallas identified four stages in creative problem solving:

1. preparation;
2. incubation;
3. illumination;
4. verification.

Preparation

This is the stage in which the facts necessary for solving the problem are gathered and initial attempts may be made to solve it. Preparation may take years or hours and often includes considerable effort searching for information, experimentation, trying to find solutions and exploring and redefining the problem. It may also rely heavily on existing knowledge and experience. Usher called this 'Setting the stage'. Other terms used are 'problem exploration', 'task clarification' or 'immersion'.

We clearly saw long preparation preceding Dunlop's idea of a pneumatic tyre. Significantly the breakthrough came from his knowledge and experience as a veterinary surgeon. Likewise John Kemp Starley, given his uncle James Starley was the inventor of the tangent spoked cycle wheel, was fully immersed in the cycling ideas and designs of his time before creating the Rover bicycle.

Incubation

At this stage the problem is set aside, deliberately or otherwise, and no longer given conscious attention. Dunlop's 20-year quest for the solution to vehicle vibration must have included many periods of incubation.

Illumination

This is the inspiration or 'flash of insight' in which the solution (or at least a key idea on which *part* of the solution might be based) suddenly occurs in the mind of the creator, often when they are not working on the problem. This is thought to be the result of the relaxed brain re-patterning information absorbed during the period of preparation, often after receiving a new piece of information that is perceived as relevant. Usher called this the 'act of insight', but he does not mention a need for incubation for this insight to occur.

There is no direct evidence that either Starley or Dunlop got their basic ideas through a major, unexpected flash of insight (a 'Eureka moment'). It may have occurred like that, but it could have been the result of conscious attempts at problem-solving resulting in a series of smaller 'illuminations'. Indeed, Mike Burrows (1993) designer of a highly innovative aerodynamic bicycle (shown in Figure 57 below) specifically said it was the result of 'no one big Eureka, but lots of little Eurekas' that occurred over a period of five years.

Verification

This final stage involves checking the validity of the idea or solution, working out the details and generally developing it for practical use. Again this may take days or years of effort without any guarantee of success. Usher called this stage 'critical revision'.

Certainly there was a fairly lengthy verification phase of further development and improvement of both the Rover bicycle and the pneumatic tyre before they were practical, commercial designs.

Although verification is usually the longest, most costly and difficult stage, more attention has been focused on the moment of inspiration than on any other stage in the creative process. Stories, such as James Watt suddenly getting his idea for a steam engine with a separate condensing chamber while out walking, highlight the fact that creative ideas may occur when someone is not consciously attempting to solve a problem. You've probably experienced an idea concerning a problem you've been working on unexpectedly popping into your mind when bathing or relaxing. But such experiences disguise the equally important fact that to get that creative idea might have involved years of thinking about the problem and trying to solve it, and may still require many years of effort to get it to work in practice.

So there is some evidence for Wallas' model. But it has weaknesses.

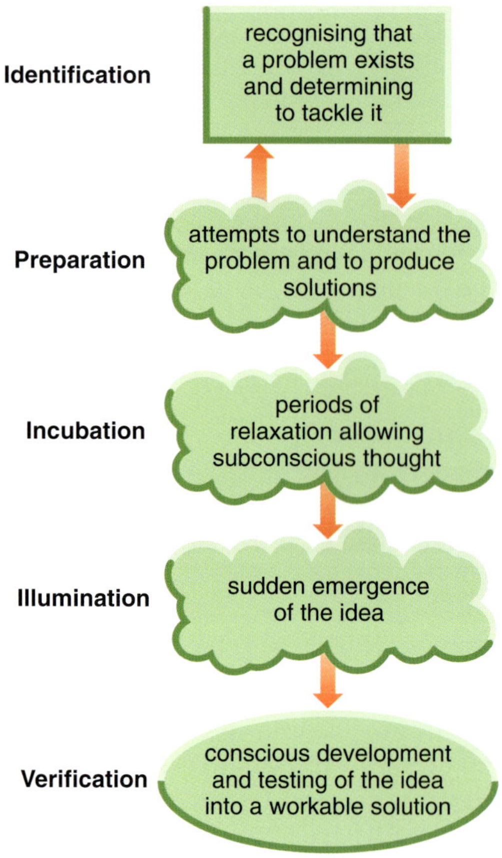

Figure 25 Five-stage model of the creative process (adapted from Lawson, 1990, p. 110)

First, it ignores the fact that creative ideas or solutions often do not occur in one unexpected big flash, but are the result of numerous smaller flashes of insight, usually when working consciously on the problem.

Second, the model fails to mention an important preliminary stage in the creative process, namely that of *problem identification*. Usher recognised the importance of this stage, which he called 'perception of the problem'.

Indeed the ability to identify problems that are worth solving is considered to be a crucial aspect of creative thinking. As you've seen, creative individuals are sensitive to deficiencies in what exists and are motivated to do something about it. They exhibit 'constructive discontent'. Starley's case shows that the identification of a problem to be solved and how it is then defined, understood or 'framed' are crucial to the nature of the solutions produced. Likewise, Burrows redefined the task of designing a faster bicycle as making the bike frame more aerodynamic, replacing previous ideas of enclosing the rider in a streamlined shell.

In his book *How Designers Think* Brian Lawson (1990) provides a five stage model of the creative process that includes Usher's initial step of problem identification (Figure 25). But Lawson warns that reality is more complicated than suggested by his simple linear model and quotes the famous engineer Barnes Wallis to argue that creative ideas often arise through extended hard work rather than a single unexpected illumination. So let us next consider some other processes that may be at work.

2.2.2 Associative thinking theories and types of associative thinking

association
A linkage or connection in the mind or imagination between two or more things. Association often, if not always, involves a process of recognition.

Many theories argue that creative thinking is a matter of forming new associations between elements of existing knowledge. These associations may arise through a variety of thinking processes, for example: scanning the memory, a systematic search for solutions, analogy between similar ideas or objects, and transfer between different areas of knowledge. This is why broad and deep *knowledge and experience* can be essential to creativity. It's also why so many of the creative problem solving techniques introduced in Section 5 of this block involve the stimulation of associations, or the production of combinations of ideas and elements.

Brian Lawson notes that before student designers can begin to produce creative ideas they need a pool of technical knowledge and experience to draw upon, but at the same time need to retain the flexibility of mind to allow that knowledge to be patterned and combined in new ways.

Let's look in more detail at different types of associative thinking. But remember that the way each works is essentially the same – a solution is provided, or an idea for a solution is prompted, by the designer making the mental association between a problem in one field and principles, knowledge or ideas from one or more other fields.

Figure 27(a) The Toggle, designed by an Open University student, combines a swivelling screwdriver and a wire-stripper to produce a new tool

Figure 27(b) Combining a fixed screwdriver with a wire-stripper creates another design of new tool using same basic concept as in (a)

Adaptation

Adaptation is probably the most common form of associative thinking. In adaptation a technical principle or solution from one field is used to provide an idea for a solution in another. You've seen the process at work in the evolution of the Rover bicycle. A more recent example is the idea for the roll-on deodorant container inspired by the ballpoint pen (Figure 26). You may be able to think of other examples. It means that one of the most fruitful sources of ideas in any field is knowledge of existing solutions, technologies, and designs in a wide variety of other fields both related and unrelated to the current design problem.

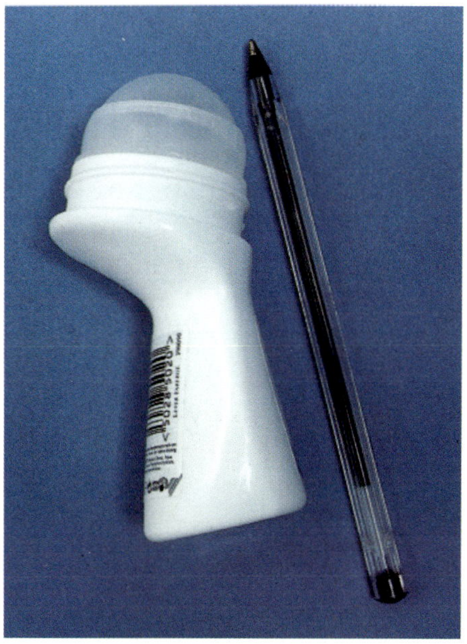

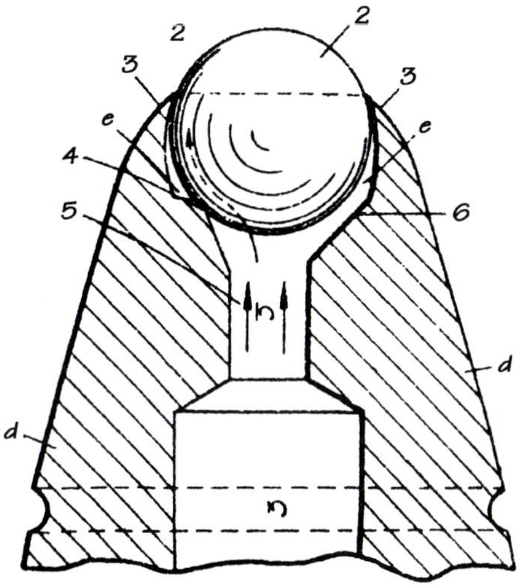

Figure 26 The idea for the roll-on deodorant container came from adapting the controlled method of dispensing ink in a ball point pen

Transfer

Transfer is a form of associative thinking similar to, but less direct than, adaptation in which the technologies, manufacturing processes or materials in one field provide the stimulus for a concept design in another. Dunlop transferred his knowledge of rubber veterinary appliances to the conception and design of the pneumatic tyre. A contemporary example is the transfer of laser technology to applications ranging from navigation and surgery to welding and sound reproduction.

Combination

Combination is a particular type of associative thinking that often produces new concepts. An elementary form is where two or more existing devices are combined to produce a new concept, which can then be embodied in different designs. An example is a combined screwdriver and wire-stripper (Figures 27a and b). A sophisticated example is a combined mobile phone, personal organiser and Internet device made possible by advanced electronics and software (Figure 28).

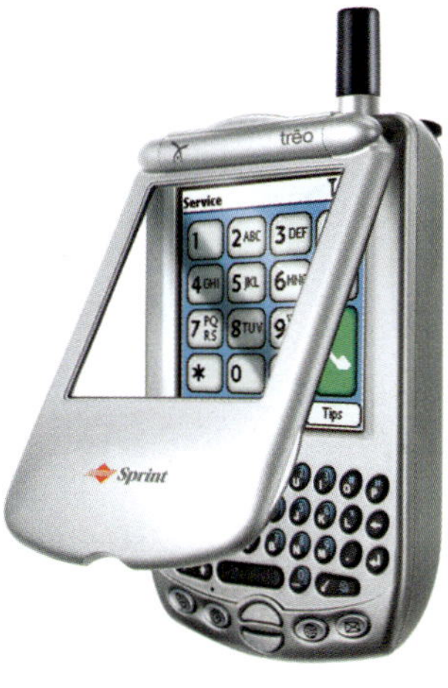

Figure 28 Handspring Treo 300 combines the functions of a mobile phone, personal organiser and computer that provides email and Internet access

Figure 29 Multifunction printer/copier/scanner/fax

Combination can also be useful for reducing the environmental impact of products. Multi-functional products such as a combined computer printer/copier/scanner/fax (Figure 29) consume less energy and materials than four separate machines. This is an example of *designing for the environment* that you met in Blocks 1 and 2, which will be considered further in Section 6 of this block.

Some creative problem solving techniques that will be outlined in Section 5 are based on combining existing ideas or objects to produce something new.

Analogy

Analogy– drawing on *similar* ideas or objects – is another type of associative thinking you've already met.

Jonathon Ive, head of industrial design at Apple Computer, often uses analogies to inspire ideas for new products. For example, the form of a transparent mouse designed for Apple came from Ive thinking about how a drop of water sits on a flat surface. The Anglepoise lamp inspired the family of white iMac computers with their processors in a dome-shaped base and fully adjustable flat screens on a jointed neck attached to the base. Ive's hugely sucessful iPod hard disk digital music player (Figure 30) was designed to be slim, desirable and pocketable 'like a cigarette pack for people addicted to music rather than tobacco' (Arlidge, 2003, p. 19).

Figure 30 Apple iPod digital music player: 'a cigarette pack for people addicted to music rather than tobacco'

Biological analogies in particular often provide or stimulate ideas for invention and design. A well-known example is the *Velcro* fastener whose nylon surfaces of hooks and loops link together. The idea occurred to a Swiss engineer, Georges de Mestral, when he saw how the tiny hooks on some seed pods stuck to the loops of wool of his clothes.

There are many other examples of plants and animals providing useful ideas and principles for design. For example, a biological analogy provided design ideas for the Kingsgate footbridge in Durham. When thinking about the main structural members, its designer Ove Arup later said that he 'saw them like a thing growing out, like a rhubarb stem'. In early sketches the concave stems of rhubarb were copied; see Figure 31(a). Later Arup made the supports convex because he realised that water might collect where two concave struts joined at the base.

Biological analogies can also provide product styling ideas, for example the stapler handle designed to look rather like an insect shell in Figure 31(b) or the Philips webcam designs inspired by animal forms (shown on the course DVD). In these cases the biological analogy provided the basic idea for the design rather than the final form.

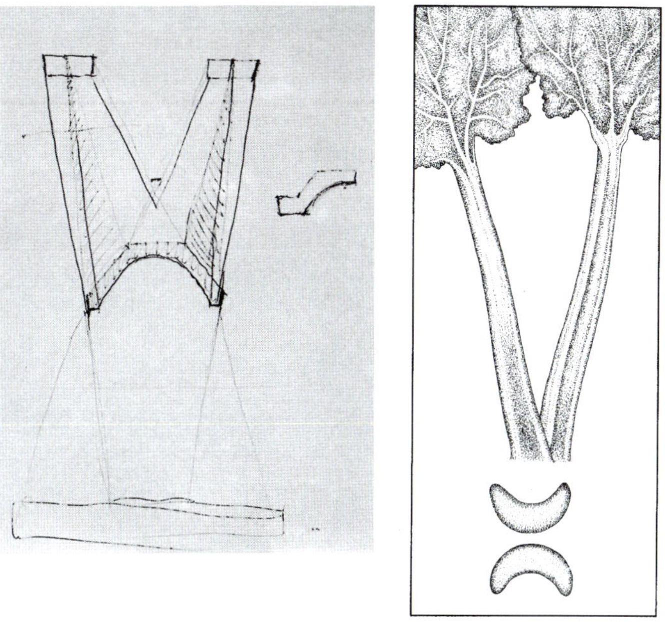

(a) A page from Ove Arup's sketchbook showing the struts of the Kingsgate footbridge (left)

Arup's initial idea for the main struts of the Kingsgate footbridge in Durham was based on an inverted rhubarb stem (right). (If you want to see details search for 'Kingsgate footbridge' on the Internet.)

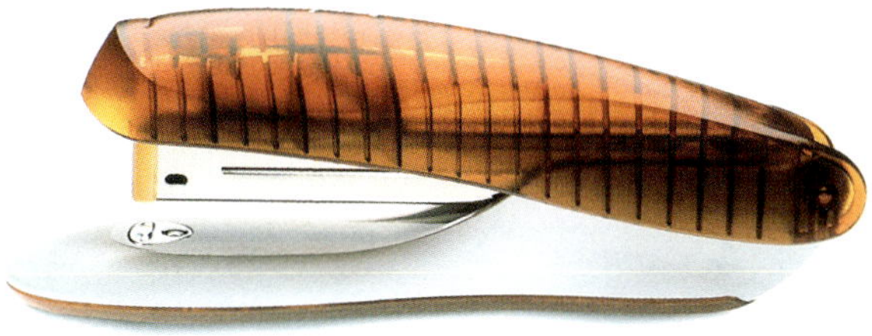

(b) Styling of a stapler inspired by insect-like forms and colours

Figure 31 Biological analogies

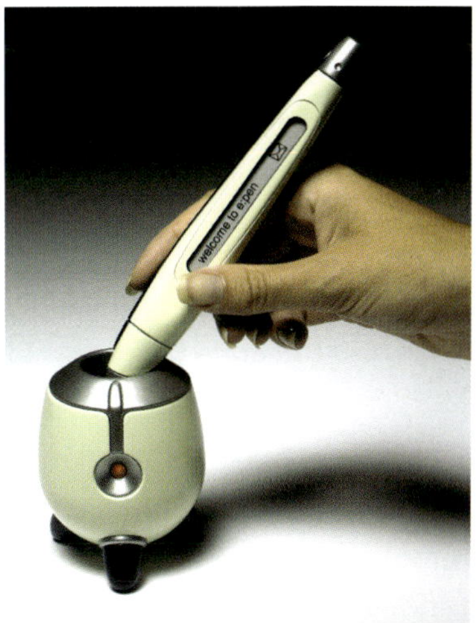

Figure 32 Concept model of an 'email pen' and 'inkwell' designed by Cambridge Consultants Ltd.

SAQ 3

What creative thinking process(es) are likely to have been involved in the following cases?

1. The concept design for a bridge based on the vertebrae of a dinosaur that its designers produced after a visit to the Natural History Museum in London.
2. A designer using a pistol grip originally designed for guns in the design of other objects from power drills to cameras.
3. A 19th century designer's dissatisfaction with the uncomfortable saddles used on cycles of the time that led him to design a bicycle with a hammock like seat.
4. The idea for an 'email pen' that enables the user to write an email, send it via a radio link to a small 'inkwell'-shaped computer which transmits it to the Internet and on to the recipient (Figure 32).

2.2.3 Divergent thinking theories

In the late 1950s and 1960s an American psychologist, J.P. Guilford argued that creative problem solving was related to a mental ability he called divergent thinking. Divergent thinking enables people to 'diverge' from what is known to produce novel ideas. He concluded that divergent thinking comprised separate abilities which he called fluency, flexibility and originality.

divergent thinking
The mental ability to produce many different or original ideas for tackling a problem.

Fluency is the ease with which individuals produce many different responses to a problem. One example of a test of what Guildford called 'ideational fluency' is 'in three minutes, write down as many uses as you can think of for a brick'. Design students on average produce between ten and twenty ideas in this exercise (you might like to try it for yourself before you read on).

Flexibility is the ability of an individual to alter their mental set and to produce ideas outside of the usual categories. Ideas such as using a brick as the weight for a pendulum show flexibility because they break away from the usual view of bricks being used for building.

Originality is the ability to generate novel, unusual or ingenious ideas such as Carl Andre's notorious rectangular arrangement of bricks as Minimalist art.

convergent thinking
Thought processes involved in selecting an idea or solution from many alternatives.

Guildford also said that another mental ability, that he called convergent thinking, was essential to creativity. Convergent thinking he argued was needed *after* divergent thinking had created many ideas, to employ logic and information in order to 'converge' on a single idea or solution.

The work of Guilford and others has been very influential in creative thinking. It is the basis of brainstorming and other creative problem solving techniques that are specifically intended to stimulate people to think of as many ideas as possible for tackling a problem. Divergent thinking also has much in common with Edward de Bono's concept of '*lateral thinking*' – ways of breaking out of familiar patterns and solutions by thinking 'laterally' or sideways (e.g. de Bono, 1977).

However, it's a mistake to equate divergent or lateral thinking with creative *design* ability. It's possible to produce lots of highly divergent ideas, none of which are feasible or have practical value. In creative design, convergent thinking (to evaluate and select ideas and concepts that satisfy a brief or specification) is as important as divergent thinking (to generate ideas). Michael French even argues that in engineering design:

> ... it would make much better sense to call creative processes convergent ... The designer is choosing one form from many ... one of the distinguishing features of the good designer is the ability to *converge from a wide base* onto a good choice.
>
> (French, 1988, p. 277)

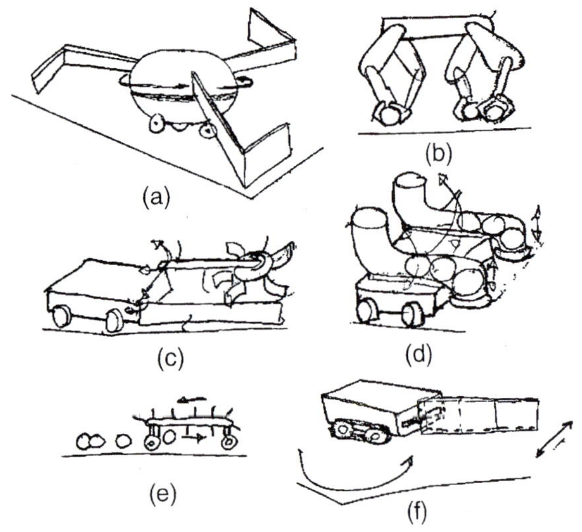

Figure 33 Concept designs for an independent device for collecting golf balls and returning them to a storage area
These divergent concepts generated by one engineering student are based on different *technical principles* of operation as well as different *configurations*.

This of course begs the question of where the 'wide base' of ideas comes from in the first place. It could be from an experienced designer's accumulated store of information, but it could stem from a divergent search for ideas. For example, Figure 33 shows divergent concepts for a golf ball-collecting device produced by a creative engineering design student (Shah and Vargas-Hernandez, 2003, p. 128). What French probably means is that a creative designer must be able to choose fruitful solutions from the possibilities they might have generated within the mind or explicitly on paper (as in the car horn case study in Section 1 or this golf ball device).

A good designer is someone who has both sets of mental skills needed to produce creative work and can *switch* between the two. At one moment they turn off their critical faculties to consider many ideas within the solution space, they then subject these ideas to rigorous evaluation and choice. This is repeated many times during the designing process. This is essentially the model of designing you met in Figure 15, but with the explore/generate (i.e. diverge) and evaluate/select (i.e. converge) cycle happening frequently at each phase. Experienced designers can converge onto ideas without explicitly considering many alternatives but, as you will see in Sections 4 and 5, learner designers can benefit from deliberately generating as wide a variety of concepts as possible before converging.

2.2.4 Solution-focused theories

The view that creative design is mainly convergent comes from studies of how experienced designers actually work. Because there are usually many solutions to a design problem, in order to make the problem manageable designers often envisage a concept design before the problem is fully defined or explored. This concept is then tested against the realities of the problem and modified as required, rather than arising (however creatively) out of an analysis of the problem. The tendency of experienced designers to start with solutions – or ideas for solutions – has been called a *solution-focused strategy*. This is in contrast to the problem-focused strategy typically adopted by scientists, who tend to analyse a problem before coming up with a solution.

What is important here is that the concepts that designers create often have their origins in an idea, image or objective arising from their experiences, preferences or design philosophy. This generating idea – or the initial solution itself – has been called the primary generator (Darke, 1979), similar to the 'problem frame' mentioned earlier.

primary generator
An idea, image or objective that generates an initial solution to a design problem; or the initial solution itself.

Bryan Lawson describes the primary generator as follows:

> A designer has some basic idea about the form the solution could take and a crude design is predicated on this basis and tested. Sometimes these tests will lead to a refinement of the idea and sometimes to its rejection, but in either case the designer learns more about the problem.
>
> (Lawson, 1994, p. 5)

The notion of the primary generator arose from studies of how architects tackle design problems. These showed that, when faced with complex design problems and numerous constraints, architects tended to latch onto an idea, image or objective early on to enable them to generate an initial solution concept. This primary generator might be concerned with practical functions such as wanting to make best use of available light or open space, or psychological objectives such as wanting to express a particular mood through a building.

Often the primary generator stems from a preferred material or geometrical form. For example, the architect Richard MacCormac said 'all our schemes have a geometrical basis … whether it is the pinwheel … grid … or the circle'. A favourite design from another context often provides the idea, like the Anglepoise lamp inspiring

Jonathon Ive's adjustable screen iMac computers. Or the generating idea might arise from social objectives, such as the 'glass eyeball' design for the Greater London Authority's headquarters attempting to express democratic openness. But such primary generators depend on the designer having a large store of knowledge retained in terms of general principles as well as specific examples and solutions. Thus Michael French says:

design repertoire
A designer's store of knowledge, comprising technical principles, examples and solutions.

> The mind of the designer should be like a rich, open soil, full of accessible resources, of which the chief is an ordered stock of ways and means, illustrated by many examples. The sum total … might be called the 'design repertoire' … the repertoire should be organised on the most abstract lines possible, so that any item … is likely to be recalled in as many contexts as possible.
>
> (French, 1988, p. 300)

Here again we see the importance of the store of knowledge that provides experienced designers with their raw material for creative ideas and concepts produced through forms of associative thinking.

2.2.5 Designing from scientific and technical principles

A designer's knowledge of scientific and technical principles is an important source of creative ideas and solutions especially in product and engineering design.

An example of designing from such principles is Gordon Murray's innovative design of the steering column of a sports car. Murray was able to redesign the conventional heavy solid steel column as a lightweight two-part system.

Look carefully at Figure 34. The conventional steering column (a) comprises linked solid steel bars supported from the dashboard to take both bending when the driver leans on it and twisting loads from steering, all encased in a plastic housing for cables and switches. In Murray's new design (b) the structures carrying each load are *separated*. Bending loads are carried by the outer housing fixed to the car, while a thin inner aluminium tube that turns within the housing carries twisting loads. Murray was able to create this novel solution by thinking from *first technical principles* about the different loads acting on the column and designing structures that could best support them (Cross, 1996).

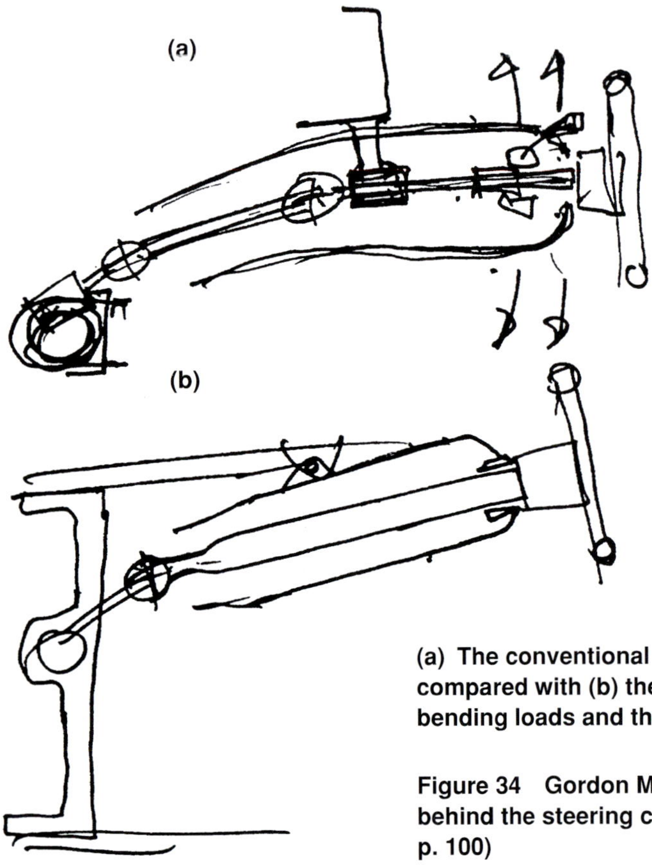

(a) The conventional solid steel steering column with plastic housing is compared with (b) the column of the F1 in which the outer housing takes bending loads and the inner tube takes twisting loads

Figure 34 Gordon Murray's explanatory sketch of the design thinking behind the steering column of the McLaren F1 sports car (Cross, 1996, p. 100)

Product and engineering designers therefore benefit by knowing as much as possible about scientific phenomena and technical principles that can provide a solution to a problem. It is estimated that an experienced designer may know about fifty such phenomena and principles. However, there are also software packages (e.g. Invention Machine's *IM Phenomenon*) containing databases of several thousand scientific phenomena and technical principles that can help designers broaden their 'design repertoire' in the search for innovative ways to solve problems.

2.2.6 Creativity as conflict resolution

As you learned from Block 1, product designs have to satisfy many requirements. These typically include function, cost, ease of manufacture, appearance, reliability and safety. And as Block 2 showed, usability and minimum life cycle environmental impacts are increasingly important requirements. These different requirements often conflict. For instance, function may conflict with appearance, or reliability with cost. Creative design may be viewed as *resolving* the many requirements with an elegant and satisfying solution. Thus, Moulton (1979) said that the design of his Hydrolastic suspension (Figure 2) did not become definitive 'until a proper harmony between the elements had been achieved'.

Another example is the sewing machine redesign problem discussed in Block 2. Its designer Kenneth Grange's goal was to follow his philosophy of creating products that are a pleasure to use. His clients' goals were to restyle the machine and to make it easier for the user to clean and maintain. Grange resolved these different requirements with an elegant design in which usability was improved with changes that also improved the machine's appearance. You may wish to refresh your memory of this design by looking at Figures 9 and 10 in Block 2.

A third example of designing as the creative resolution of conflicts comes from the renowned Finnish architect Alvar Aalto, whose work appears in Block 1. Aalto has written:

> I forget the whole mass of problems for a while, after the atmosphere of the job and the innumerable different requirements have sunk into my sub-conscious ... I just draw by instinct ... and in this way, the main idea takes shape ... which helps me bring the contradictory component parts into harmony.
>
> (quoted in Irvine, 1992, p. 60)

Here you see again the stages of preparation and incubation, but with the mental resolution of conflicting requirements coming through *drawing* rather than through an unexpected illumination. Aalto says he draws 'by instinct' but his sketches are actually based on vast experience (including his design repertoire). I'll return to the importance of drawing in creative design in Section 5.

2.3 Creativity and innovation

You've seen some of the ways in which the basic idea underlying a new design may arise. However, creating this basic idea and embodying it in concept designs is often the easiest part. Development of the concept and making it into a workable solution is the hard grind of *verification* (recall the five stage creative process model in Figure 25). The verification process will normally throw up further problems that require creative solution.

Case study The 'Cyclone' cleaner

The effort involved in developing a novel concept into a marketable product is dramatically illustrated by the Cyclone cleaner. James Dyson got the basic idea for his cyclonic cleaner by *transfer* from an industrial cyclone. After rapidly testing the idea with an experimental mock-up (principle-proving model), he spent several years developing his Cyclone cleaner by building and testing over 5000 model cyclones and several prototypes (Figure 35). Dyson's initial development work was largely based on trial-and-error experimentation, a very laborious process.

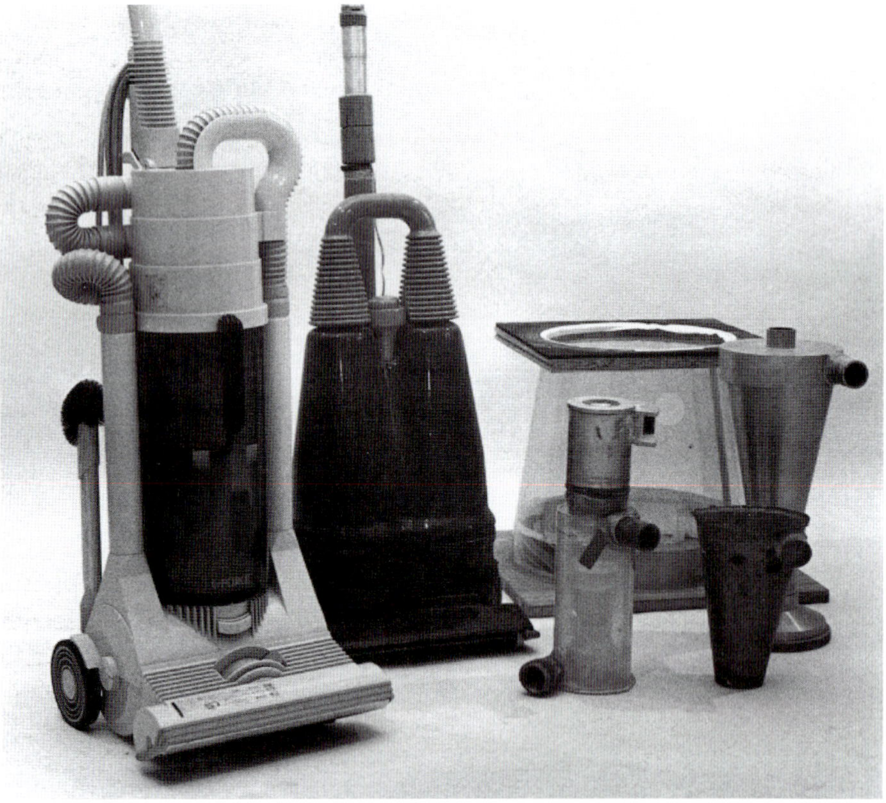

Figure 35 The G-Force cleaner (left) with concentric cyclones, made in Japan; Dyson's first prototype (centre) in which the cyclones were placed side by side; some of the thousands of models (right) used to develop the best shape of cyclone

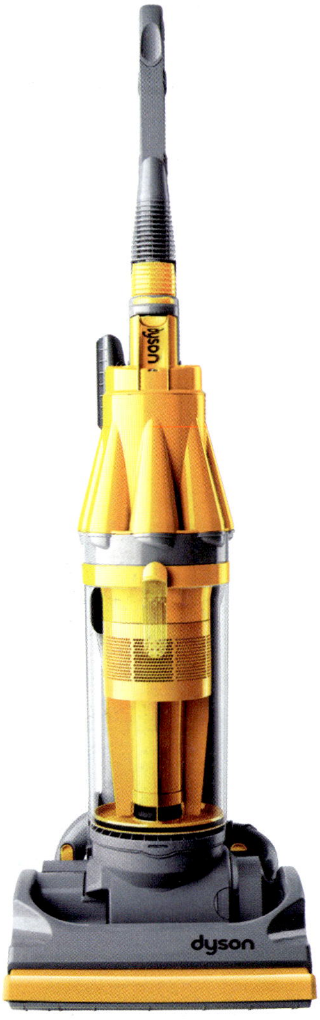

Figure 36 Dyson DC07 Animal 'root cyclone' upright cleaner with 8 small cyclones

Less laborious methods employ mathematical or computer modelling based on a theory, or a combination of theory and experiment. For example, once the cyclonic technical principle had been established and the first versions of the cleaner launched, Dyson's company went on to develop new designs with improved features and cyclonic technology (Figure 36). The modelling involved in improving the technology and creating the configurations of components for a new cylinder model, the Dyson DC05, will be discussed in Block 6.

Before you begin the next section, look at the DVD notes in the *Block 3 Supplement* and view the DVD sequence 'James Dyson: designer and innovator'. On this sequence Dyson talks about creating innovative products for the market. Using the examples of the Ballbarrow (Figure 37) and Cyclone cleaner, Dyson demonstrates his use of physical and other modelling to develop a novel concept into a working prototype and to create designs that improve upon existing products. After viewing, listen to the Course Team comments on the DVD and attempt the following self-assessment questions. You may need to watch the sequence again when studying Section 6 of this block.

Figure 37 James Dyson (right) discussing his 'Ballbarrow' with the author of this block

The design arose from Dyson's dissatisfaction with conventional wheelbarrows. The idea for a ball-shaped wheel to improve the ride over rough ground came from Dyson's experience of rotational moulding of plastics.

SAQ 4

What types of modelling were used in the development of the first Cyclone cleaner? What models were used subsequently to develop new improved designs?

SAQ 5

Distinguish between the technical principle, the design configurations and the design features of the Cyclone cleaner.

SAQ 6

According to Dyson, how expert do you have to be in a particular technical field to produce worthwhile innovations?

SAQ 7

Why does Dyson consider that practical work with real physical objects is necessary for creative design and technical innovation? What other approaches are commonly used?

SAQ 8

On the basis of what you have learned so far, do you consider that Edison's observation: 'genius is 1% inspiration and 99% perspiration' is valid?

Before you read on, look over this section and in your workfile make a summary of its key points in six or more bullet points. Do this before looking at my own summary below, then compare your key points with mine.

Key points of Section 2

Creative individuals are dissatisfied with what exists and wish to do something about it – '*constructive discontent*'.

In many cases the general creative process goes through five stages (Figure 25).

1 *Identification* – recognising a problem exists and determining to tackle it.

2 *Preparation* – attempts to understand the problem and to produce solutions.

3 *Incubation* – periods of relaxation allowing subconscious thought.

4 *Illumination* – spontaneous emergence of a breakthrough idea. More often this may involve a series of smaller 'illuminations'.

5 *Verification* – conscious development and testing of the idea into a workable solution.

Creative design ideas, and inventions, in all fields frequently arise through *associative* thinking, including:

direct *adaptation* of a technical principle or solution from one field to provide an idea for a solution in another;

indirect *transfer* of the technologies, manufacturing processes or materials in one field to provide the stimulus for a concept design in another;

analogy with similar objects or situations, including biological objects;

combination of pre-existing ideas or solutions to produce something new.

Experienced designers often follow a *solution-focused strategy*. They have an idea for a solution early on and work this into an initial concept design in order to explore the problem and develop a better solution.

The experienced designer's initial idea or solution arises either from an image or objective arising from their experiences, preferences or philosophy – the *primary generator* – and/or from their particular view or understanding of the problem – the '*problem frame*'.

A creative solution may arise by *designing from first scientific/technical principles*. The problem frame may help the designer to identify relevant principles.

Creative design often involves *resolving many, often conflicting, requirements* with an elegant and satisfying solution.

A good designer is someone who is capable of, and knows when to switch between, *divergent* and *convergent thinking*. While *experienced* designers can be solution-focused and quickly converge on an initial concept design, *learner* designers can benefit from deliberately generating *a wide variety of concepts* before converging.

For designs involving a new technical principle, practical *experimentation* with a principle-proving model or prototype is required to test basic feasibility. Such experimentation is not essential for designs that are new configurations of existing technology, although it may produce creative design ideas.

(Learning outcomes 1.2–1.4, 2.1 and 2.2)

3 Creative designing – individual approaches

In this section I shall first look at four contemporary examples of how creative individuals designed innovative products, again in the field of bicycles. I then draw upon these examples to obtain a fuller understanding of the creative process, especially as it applies to concept design.

Figure 38 Alex Moulton riding his small-wheel bicycle

3.1 Bicycles – contemporary examples of creativity

3.1.1 The Moulton bicycles

Early pedal cycles had large wheels due to the lack of gears. The driving wheel of the 19th century Ordinary bicycle had a diameter of four to five feet and, even though it had gearing, the first Rover bicycle was equipped with three-foot diameter wheels. By the end of the 19th century, following the introduction of the pneumatic tyre, the diameter of wheels had been standardised at between 26 and 28 inches, a size that gave a comfortable ride on ordinary roads and fitted nicely into the adult diamond-frame configuration.

The idea of a small-wheel bicycle for adults is not new; a prototype small-wheeler was built in Paris in 1868. Although adult small-wheel bicycles are now commonplace, it was not until 1959 that a serious attempt was made to design one. This was the bicycle created by Alex Moulton, with sixteen-inch diameter wheels and a rubber suspension system, which was launched onto the British market in 1962 (Figure 38).

Case study The Moulton small wheel bicycle

The significance of the Moulton bicycle was not just its innovative design, but the fashionable image and revival of interest it gave to cycling. It also stimulated new cycle design that until then had largely converged on the classic diamond frame bicycle – the dominant design.

dominant design
After a period of technical change, a design configuration that has largely displaced competing designs and so has become the accepted market standard

As with most highly creative work, the context is significant. Alex Moulton trained as a mechanical engineer at Cambridge University, and worked in the aircraft industry before joining his family firm. The firm made rubber springs, and Moulton had the idea of developing rubber suspension springs for road vehicles. I noted earlier that, in collaboration with Alec Issigonis, Moulton patented and developed the rubber cone suspension system first used in 1959 on the Mini car, and he subsequently developed the Hydrolastic suspension system used from 1962 onwards on the BMC 1100/1300 series cars (Figure 2).

At the same time that he was developing the Mini suspension system, Moulton's boyhood interest in cycling was revived as a result of the petrol shortages caused by the Suez crisis. He gave the following account of how he came to develop his small-wheel bicycle.

> There is an exercise for you to do following the extracts from Moulton's account in the case study. You might find it helpful to briefly look at Exercise 6 before reading his account.

Moulton's account

It was the Suez crisis in 1956 which spawned the Mini as the economical car, and got me interested again in cycling. ... I took a serious interest in the machine itself, by using one ... under the spur of the petrol shortage. I pondered on what it comprised and how it had evolved to what it was. Having at that time established the laboratory for the car suspension development I had some facilities for making design studies leading to actual hardware prototypes.

'... "Good ideas" are two a penny but it is hammering them into the harsh reality of a working device that is the difficult thing ...'

The question of the minimum 'kit of tools' for any innovatory achievement is an important one. 'Good ideas' are two a penny but it is hammering them into the harsh reality of a working device that is the difficult thing, and this demands some sort of making facility. ...

'The first thing, as always, was to make a study of the fundamentals ...'

I resolved therefore in 1959 to 'have a go' at improving the design of the bicycle — to make it more pleasing to use and possess. The first thing, as always, was to make a study of the fundamentals. I questioned the riding position of the classic machine and acquired a pre-war Grubb to carry out tests of the alternative reclining [recumbent] position. Although good for a short period of activity it was tiring for the longer exercises. I confirmed, therefore, the virtues of the 'classic' position.

[...]

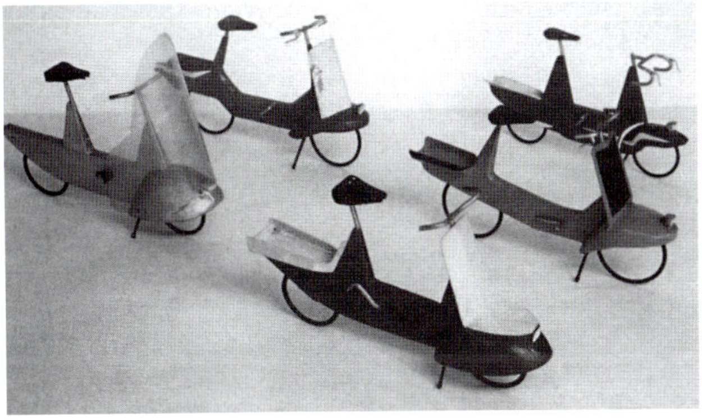

Figure 39 Scale models made prior to construction of the first prototype Moulton bicycle

Figure 40 First prototype Moulton bicycle with light alloy sheet, monocoque construction

I questioned also the size of the wheel: remembering that in all other vehicles the reduction in size of wheel has been in the direction of design evolution because so many advantages generally speaking flow from it. We were just entering the start of the Mini era with its small wheels and my rubber suspension, and it was not difficult for me to get designed and made by Dunlop a 14 inch and 16 inch wheel and tyre with a 'proper' 1 3/8 section. ... This is an example of carrying out the 'key experiment' properly: and here is the first full-size prototype [Figure 40]. You will see the 'open' frame [i.e. no cross-bar], which was the innovation of the unisex principle, and the carrying capacity ... set low down due to the small wheels. I had feared the effect of the reduction in 'flywheel effect' in reducing the size of the wheel. In all engineering innovations ... the originator will have gone through agonies of doubts. Certainly I did over this. In fact the balancing function of a bicycle is highly complex: and there are advantages as well as disadvantages of small wheel size, and in the event the behaviour was totally accepted. What was apparent was the necessity of wheel suspension to make the small diameter narrow section tyres acceptable from shock transmissions. [...]

I was quite wrong in my original choice of the structure. Coming as I had from aeronautical origins by way of contact with the motor car it was natural for me to choose a light alloy sheet metal construction [Figure 40]. It was immediately apparent in the reality of the testing that the high concentration of loads in a bicycle demanded a 'boney' structure. Moreover, the 'boxy' sheet metal frame introduced ... road-excited noise.

Then quickly followed a period of development of the optimised construction [Figures 41 and 42]. When a problem is accurately enough posed then finding the solution can be quick but it is important to scan all possible routes to ensure optimisation.'

(Moulton, 1979, pp. 40–41)

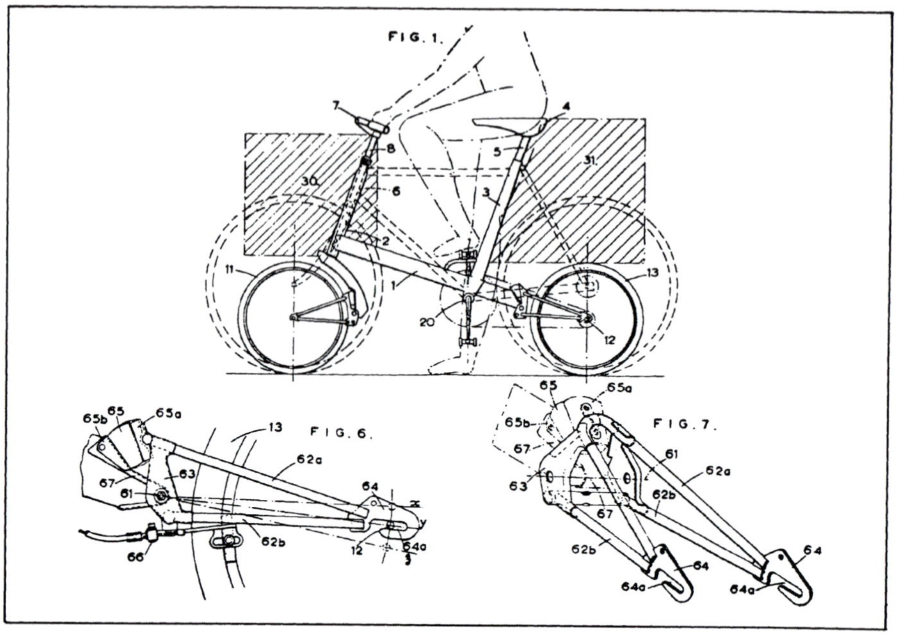

Figure 41 Original patent drawing of the Moulton bicycle showing the uni-sex frame and large load carrying capacity

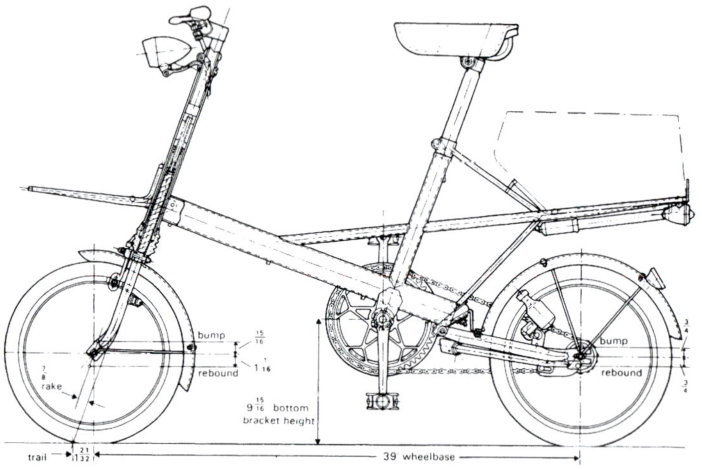

Figure 42 Engineering drawing of the Series One Moulton bicycle showing its general arrangement

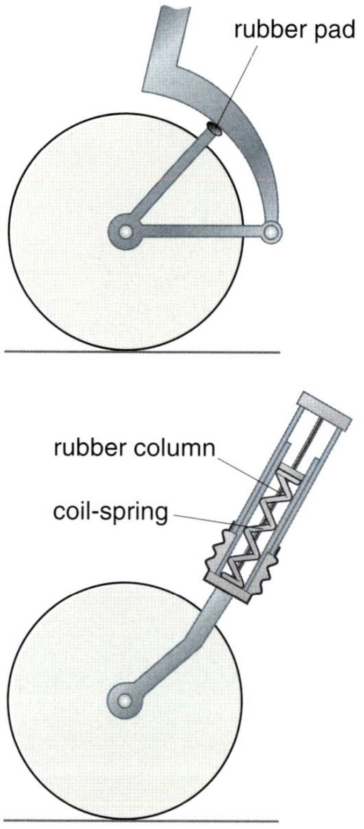

Figure 43 Two stages in the evolution of the Moulton front suspension system

Apart from the small wheels and the 'unisex' frame, a key feature of the Moulton bicycle was the rubber suspension system. Moulton considered that suspension was essential for rider comfort because small wheels with high pressure tyres are less able to absorb road shocks and will fall deeper into holes than would a large wheel. Given his experience with rubber vehicle suspensions, it's perhaps not surprising that Moulton's bicycle ended up with rubber springing. This suspension system went through several stages of development before the final version emerged (Figure 43). The final design involved a telescopic coil-spring and rubber front suspension and a trailing-link rear suspension also sprung by rubber. This shows that it is often necessary for a designer to produce and test several concepts for the key components of an innovative product.

Exercise 6 Lessons of the Moulton bicycle

On the basis of the theories of creative designing you have learned in the block so far, what do you consider to be the most significant points about the processes that led to the conception and design of the Moulton bicycle?

How does it compare with the creative process involved in J.K. Starley's design of the Rover bicycle and Dyson's Cyclone cleaner?

Spend 10 minutes making notes on your answer before looking at my response below.

My response

Below is my analysis of Moulton's creative process based on the concepts you have met so far, plus some new ones.

1 Moulton was an experienced engineer who was dissatisfied with existing products and was determined to improve upon them. ('I resolved … to "have a go" at improving the design of the bicycle – to make it more pleasing to use and possess'.) Like Starley and Dyson, Moulton displayed *constructive discontent*.

2 Like Starley and Dyson, Moulton questioned the fundamentals of the existing *dominant design*, in this case the riding position and wheel size of the classic bicycle. ('The first thing, as always, was to make a study of the fundamentals.'). As you've seen, new engineering products sometimes arise by *designing from first technical principles*.

3 Moulton considered that the general reduction over time in the size of vehicle wheels had advantages. ('… reduction in size of wheel has been in the direction of design evolution because so many advantages generally speaking flow from it.') Smaller wheels were Moulton's particular viewpoint on vehicle design arising from his experience and preferences – his *primary generator*. In this case it led to the *basic idea* of a small wheel bicycle with advantages of unisex use and better load carrying. (The primary generator is similar to Starley's *problem frame* of optimising the rider's effort in the design of the Rover.)

4 Like Starley and Dyson, Moulton considered that in technical innovation it is essential to test and develop ideas using physical models. ('…"Good ideas" are two a penny but it is hammering them into the harsh reality of a working device that is the difficult thing.') After making sketches and scale models (Figure 39) embodying his basic idea of a bicycle using small wheels with high-pressure tyres, Moulton tested the concept using a principle proving model, in this case a full-size experimental *prototype*. This 'key experiment' also resolved his worries about the effect of small wheels on the bicycle's balance ('In all engineering innovations … the originator will have gone through agonies of doubts. Certainly I did over this.')

5 Moulton's previous experience strongly influenced his ideas. His *transfer* of rubber suspension from cars to bicycles is one example. Another is his idea of using alloy sheet construction for the first prototype. ('Coming as I had from aeronautical origins by way of contact with the motor car it was natural for me to choose a light alloy sheet metal construction.') By testing the prototype Moulton soon discovered his mistake in the latter idea and for the second prototype he made a 'jerk change' to a tubular F frame configuration which led to the final design.

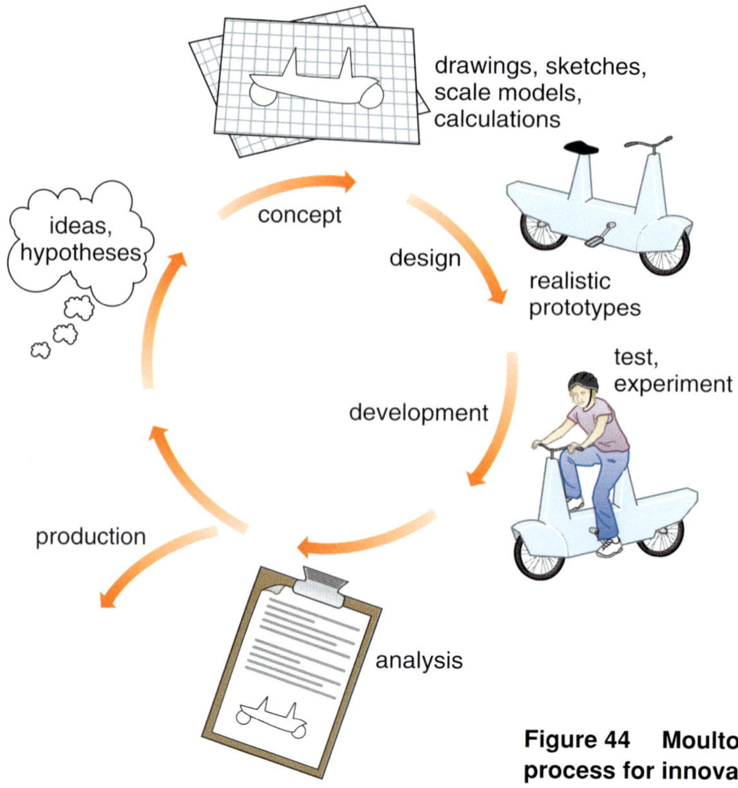

Figure 44 Moulton's creative process for innovative design

Moulton's process of creative design may be represented as shown in Figure 44. It starts with *ideas* in the mind, which are quickly embodied in *sketches* and *models* and tested through *calculations* and *prototypes*. He employed the heuristic approach of *iteration* via design/make/test, repeated as many times as necessary to optimise a final prototype. He made and tested 15 prototypes of his bicycle before the launch of the production model. Moulton clearly demonstrates a characteristic of creative individuals great *motivation* and *perseverance* in tackling a problem.

However, one thing seems to distinguish Moulton, as a qualified engineer, from his predecessors in the field of bicycle design. His ideas are based not just on experience and craft skill but on *technical analysis*. But he emphasises the need to test and analyse theoretical ideas to avoid errors – like his first prototype:

> Ideas and calculations must be translated into drawings and sketches … drawings must be made into hardware as soon as possible, so that reality can be tested and analysed.
>
> (quoted in Whitfield, 1975, p. 175)

> For innovation, I advocate the 'rapid response method of prototyping' … sketching for form and function, calculating the fundamentals, drawing to scale, and iteratively making and testing the elements within the … design of the whole. The computer for dynamic simulation is useful.'
>
> (Moulton, 1997, p. 13)

And in another lecture he firmly aligns himself with those who stress the need for *specialist knowledge* in creative engineering design:

> What differentiates the designer, who successfully innovates, from the crackpot inventor is the depth of study. I have made […] dozens of 'inventions' leading to patents; but they all arise from a revelation emanating from observing and studying in a particular field; never from a random idea occurring in a random field.

'What differentiates the designer, who successfully innovates, from the crackpot inventor is the depth of study …'

Like most innovators in an established field, Moulton encountered great resistance from the existing cycle industry to his design. To get his product onto the market Moulton had to set up a firm to manufacture it himself. However, when his small-wheel bicycles proved to be fashionable and successful, the concept was copied in the form of low-priced 'shopper' bikes with balloon tyres instead of suspension. Eventually competition from such products forced Moulton to sell his cycle firm to TI-Raleigh in 1967.

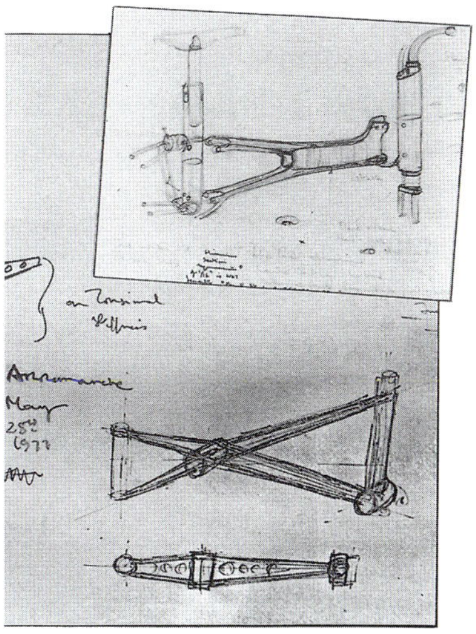

(a) Moulton's sketches of the Y and X-frames that led to the multi-tubular space frame design

Continual redesign

By 1974 Raleigh had stopped production of Moulton's original bicycles. By then he had confirmed, through laboratory tests, the lower rolling resistance of a small wheel bicycle with high-pressure tyres over a standard bicycle and started to further improve his design. In 1977 he decided to 'go it alone' once again with a new machine. The aim was to produce a bicycle that was as light as a good conventional sports bike with the advantages of his original small-wheel design. As before, Moulton created and sketched several concept designs, analysed them and then made and tested prototypes of both frame and suspension system.

In evolving the frame Moulton started with a Y-shaped configuration. This was too heavy, so he considered large diameter thin-walled tubes in aluminium or steel. An aluminium frame would eventually suffer fatigue failure at its joints and thin steel tubes dented easily, so he made the jump to an X-frame (cruciform) configuration (Figure 45(a)). This concept was developed into a multi-tubular space frame design that, as well as being light and strong, is very stiff.

> The importance of strength and stiffness in cycle frame design will be discussed in Section 3.1.3

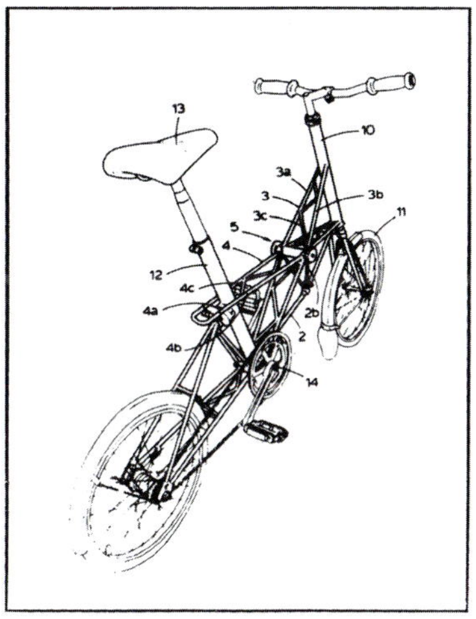

(b) US patent (1985) of the Moulton AM series bicycle showing the space frame

In his lecture Moulton noted the need for *divergent thinking* in design ('it is important to scan all possible routes to ensure optimization'). Creating the new bicycle required such divergent thinking during which many alternative concepts were explored.

The 'advanced engineering' Moulton AM series was launched in 1983, aimed at the discerning cyclist at the top end of the market (Figure 45(b)). Since then several different models have been designed, including the Moulton Speed for road racing and a version suited to off-road riding. Like many creative individuals Moulton is never satisfied. He has continued to develop new suspension designs, including a Hydrolastic rear suspension system: an *adaptation* of ideas originally used in his fluid damped car suspensions. Figure 45(c) shows one of the range of new Moulton bicycles.

(c) Moulton Dual Drive 24 speed bicycle

This particular model has 17-inch diameter wheels, leading link front suspension and rubber cone rear suspension. The range is aimed at the discerning cyclist with models priced (in 2003) from £695 for the basic model to £4995 for the stainless steel New Series Speed with Hydrolastic rear suspension.

Figure 45

3.1.2 The challenge of the folding bicycle

To increase their portability some Moulton bicycles can be separated into two parts at a central kingpin. However, the *folding* bicycle is a much more common approach to portability and has provided a challenge to designers since the 1880s. There have been numerous attempts at designing folding bicycles (Figure 46 shows a small selection). Each design, however, offers a different compromise between the conflicting requirements of a portable bicycle – folded size, ease of folding, riding performance, comfort, weight, cost, appearance. None is ideal for all purposes and markets, and many designs have never gone beyond the prototype stage. In this section I'll look at two contemporary attempts at producing a folding bicycle, to further improve your understanding of the creative designing process, including the need for technical and economic decisions on mechanics, materials and manufacture.

(a) Micro folding bicycle originally launched in 1976

To fold: frame hinges in the middle.

(b) German 'Birdy' small wheel folder

To fold: rear wheel hinges under frame, front wheel hinges sideways, and handlebars fold down.

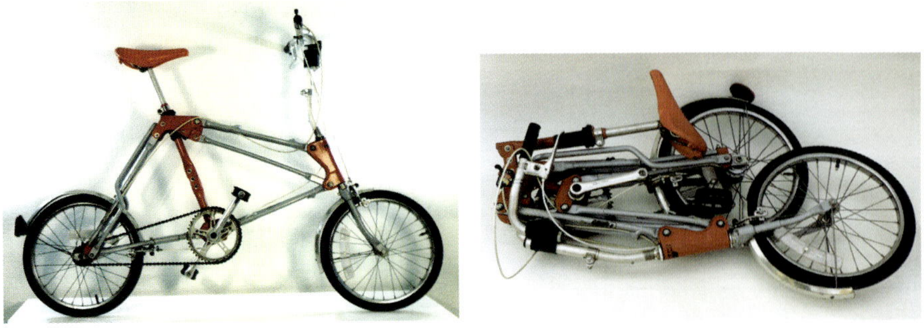

(c) 'Airframe' folding bicycle originally designed in 1979

To fold: frame concertinas together, saddle hinges down

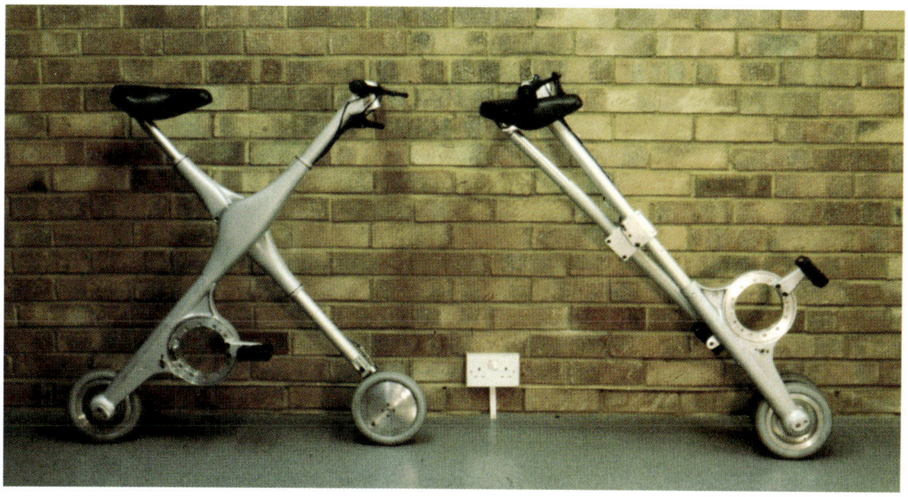

(d) Concept model and prototype Sinclair 'X-bike' with fibre reinforced plastic frame and 8 inch wheels

To fold: frame folds like scissors.

(e) Hungarian 'Taby' folding bicycle

To fold: wheels hinge into box frame which then hinges together.

(f) Prototype aluminium alloy and plastics folder that won a 1979 'Bike of the Future' competition

To fold: Handlebars and seat are lowered and wheels hinge side by side.

Figure 46 Some of the dozens of folding bicycle designs each with different folding mechanisms

The 'Brompton' is one of the most successful folding bicycles so far designed. It was created by Andrew Ritchie, a Cambridge educated engineer turned landscape designer. The account below of how it was developed is adapted from material originally written by Adrian Demaid for an introductory OU engineering course. I have revised the case study using additional information and have shown in italics the key elements of the creative designing process you've met so far.

You might find it helpful to briefly look at Exercise 7 before reading the case study.

Case study The 'Brompton' folding cycle

Bicycle origami

Andrew Ritchie started designing a folding bicycle in 1975, stimulated by the Bickerton folding bicycle. The Bickerton (Figure 47) is made from aluminium alloy to minimise weight, but this material, plus its design with long, thin handlebars and seat post, means that it flexes considerably when ridden and requires careful use and handling if it is to last.

Ritchie was inspired by the thought that he could do better (that is by *constructive discontent*). He was driven by a search for a bicycle design that when folded was 'the ultimate in compactness'. His design philosophy, which remained constant through the development of prototypes, was 'to bring the handlebars, seat, front and rear wheels to one spot' (Ritchie, 2001) (*the problem frame*).

Ritchie felt that the simplest way to 'bring together the disparate parts' was to put a separate hinge between the main frame and each of the wheels – *the basic idea*.

(a) unfolded

(b) folded

Figure 47 The 'Bickerton' bicycle, a source of inspiration for the Brompton

Ritchie realised his idea as a *concept design* by making sketches and paper models with drawing pins to represent the hinges. He rapidly proceeded to make a working prototype using standard parts for wheels, etc. 'You can only conceive so much on paper, once you have the hardware it's a great leap.'

Prototyping and improving

In Ritchie's first prototype design the rear wheel hinged forward in its own plane – as in the production model (similar to the patent drawing in Figure 48). However, unlike the production model, the front wheel of Prototype 1 also moved (almost) in its own plane underneath the bicycle to sit alongside the back one. In this case some sideways movement was needed to ensure that the front wheel sat next to the rear one, rather than catching against it as it folded. To achieve this, the front wheel needed a complex, skewed hinge to move it sideways so as to clear the rear wheel and chain wheel.

Ritchie is a regular bicycle commuter so he tests designs and design changes routinely and expertly. He was pleased with the realisation of the basic concept design in the first prototype: 'I had demonstrated that the concept design could result in a compact folding bicycle.'

The second prototype

The major design difference between the first and subsequent prototypes was the removal of the complex skewed hinge. The front wheel now hinged *sideways* from the line of the frame, as happens in the production model

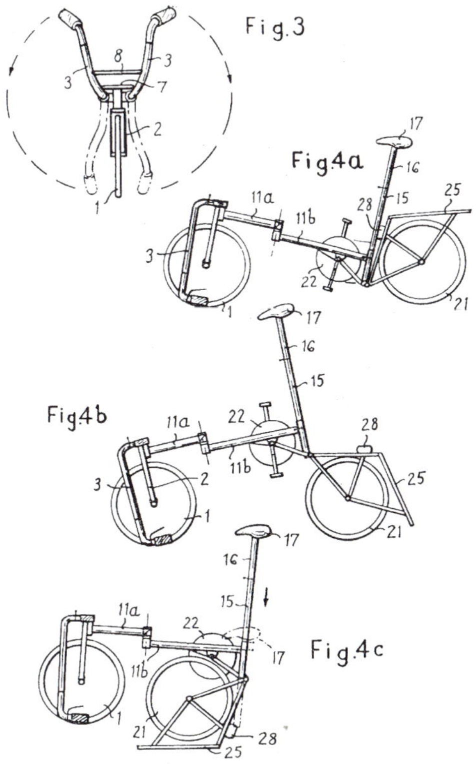

Figure 48 Drawing of folding concept of Brompton bicycle from the patent specification filed May 1977 and published November 1980

The production version is similar, although the handlebars now fold to the outside of the front wheel as in Figure 49(c).

using a purpose-designed hinge made from tubing. Ritchie's 'Eureka moment' (*illumination*) was 'realising that a perfectly satisfactory way of getting the front wheel to come back was to have a vertical pivot point' (as in Figures 48 and 49(a)).

The rear wheel continued to be folded underneath the frame, but Prototype 2's handlebars hinged *forwards* and down, one each side of the package (as in the patent drawing in Figure 48). In total Ritchie built four prototype machines between 1975 and 1979, to prove and develop his ideas (design/make/test/modify *iteration*). Another improvement was Ritchie hinging the handlebars to fold *sideways* and down to the outside of the front wheel, as in the production model shown in Figure 49. It's possible that by this time Ritchie might have seen and *adapted* similar mechanisms used on an imported folding bicycle, the De Blasi. (These details are only included to show how the design evolved. You don't have to remember them.) As he developed the design, Ritchie realised that a useful bonus was that folding the wheels one on each side of the chain wheel would protect the user from the oily chain and chain wheel – an advantage over other designs like the Bickerton.

His next task was to turn the design into a product that you or I could buy.

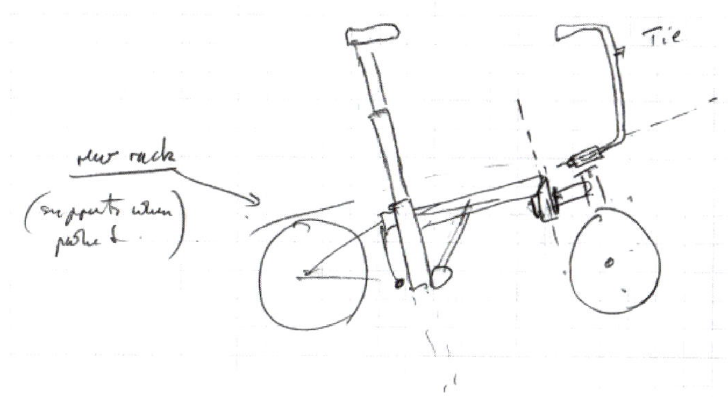

(a) Sketch of production model

(b) Andrew Ritchie with part folded Brompton – rear wheel folds under

(c) Fully folded Brompton – front wheel hinges towards rear wheel, handlebars fold over, seat post pushes down

(d) An early advertising poster for the Brompton

Figure 49

There is a complete sequence of Ritchie folding the Brompton on the DVD.

The first production run

The next part of the Brompton story shows that as much creativity may be needed at the detail stage, in which the product is designed for manufacture, as at the concept stage of design.

Andrew Ritchie could not sell his idea to any established manufacturer. So, in 1981, he raised the money to set up a small factory to manufacture the bicycles himself. After the complications of the prototyping stage, manufacturing constraints become a powerful influence.

Ritchie had decided to make his bicycle from steel because, although heavier than the aluminium alloy used for the Bickerton, it offers many advantages. Ritchie felt 'aluminium is too soft for a folding bicycle, it just doesn't stand up to the knocks, the everyday wear and tear'. More important for his small firm, steel could be fabricated using simple tools and is less likely than aluminium to suffer fatigue failure in extended use, thus avoiding possible product liability claims.

In the production design detailed changes were made, for example to the hinging system where a forged offset hinge from France replaced the complex purpose-built tube hinges.

Small-scale production

Designing and manufacturing low-cost tooling for production of the bicycle was a harder job for Ritchie than designing the bicycle. He had to use mild steel for the main tube as he could not bend stronger alloy steel and it had an aesthetically ugly kink from the bending process. Plastic parts were machined from solid blocks and metal plate was drilled, cut and bent. The process was essentially scaled-up craft production.

Mass production

Later in 1981, after about 300 Bromptons had been made, the French manufacturer of forged hinges discontinued production. This soon forced Ritchie to stop his batch production and to consider how to continue making the Brompton. He developed a business plan and, by 1986, Ritchie had raised £90 000 – half of the money he needed to go into mass production – and went ahead anyway.

The drive to design for speed and economy of manufacture continues. A tool was designed to curve the main tube, so removing the ugly kink, and a power press allowed the use of a higher specification of steel for tube. The hinges were machined from the solid steel until expensive forging tools could be bought in 1987.

You can buy a Brompton bicycle for between £400 and £600. Andrew Ritchie won a Queen's Award for Export Achievement in 1995 when his turnover was about £2 million, 60% exported. Not bad for a company that, literally, started under the railway arches.

(Adapted from Open University, 2001, pp. 52–70 and Ritchie, 2001)

Exercise 7 Lessons of the Brompton bicycle

The design of the Brompton shows the familiar creative elements of *constructive discontent*, a *problem frame* and a *basic idea*, leading to a *concept design* developed through *modelling* and *prototyping*. The approach was design/make/test *iteration*. There were moments of *illumination*.

What additional elements of the creative design process can you find in the account of the Brompton's development?

Spend 5–10 minutes making notes in answer to this exercise before reading my response below.

My response

A new element is the importance of *choice of materials* in design. As you saw in Block 1, the form and manufacture of products such as chairs can be driven by materials choice. Ritchie rejected aluminium alloy, as used for the Bickerton. Steel could provide the required strength and stiffness using relatively small diameter tubing. It also offered the advantages of manufacture with low cost tools and fatigue resistance. In the case of the Brompton this choice was made without the detailed *technical analysis* that often characterises more complex design problems. The *compromise* involved is the weight (about 2 kg more than the aluminium Bickerton).

The Brompton also indicates the considerable creativity required to design a product and its production process for *ease and economy of manufacture*. This was especially critical for the Brompton given the severe limits on investment for tooling available to such a small firm.

The bicycles discussed so far illustrate an heuristic approach to creative designing by individuals. The next example, presented on the DVD, shows a rather more *systematic* approach, although still by an individual designer.

Case study The 'Strida' folding bicycle

The 'Strida', another folding bicycle, was designed by Mark Sanders while a postgraduate student in industrial design engineering (Figures 50, 51 and 52). With only a single gear and simple triangular frame, the Strida was intended for short trips in conjunction with other modes of transport, to be low cost and modern-looking in design, and very quick and easy to fold. Sanders has improved the design through Strida 2 and 3 versions (Figure 53) in response to user feedback. Since first launched in 1987 it has sold well, but in lower volumes than the Brompton, which given its three- or five-gears is suited to both short and longer trips.

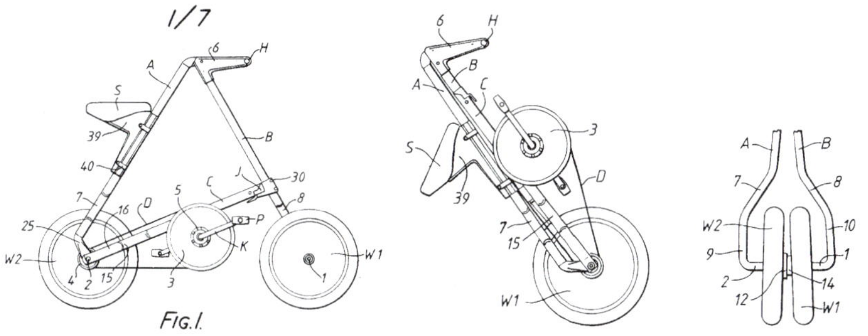

Figure 50 The original Strida patent and a folded Maclaren baby buggy that helped inspire it

This shows the triangular frame designed to fold into a long thin form that could be easily wheeled along the ground like a folded baby buggy.

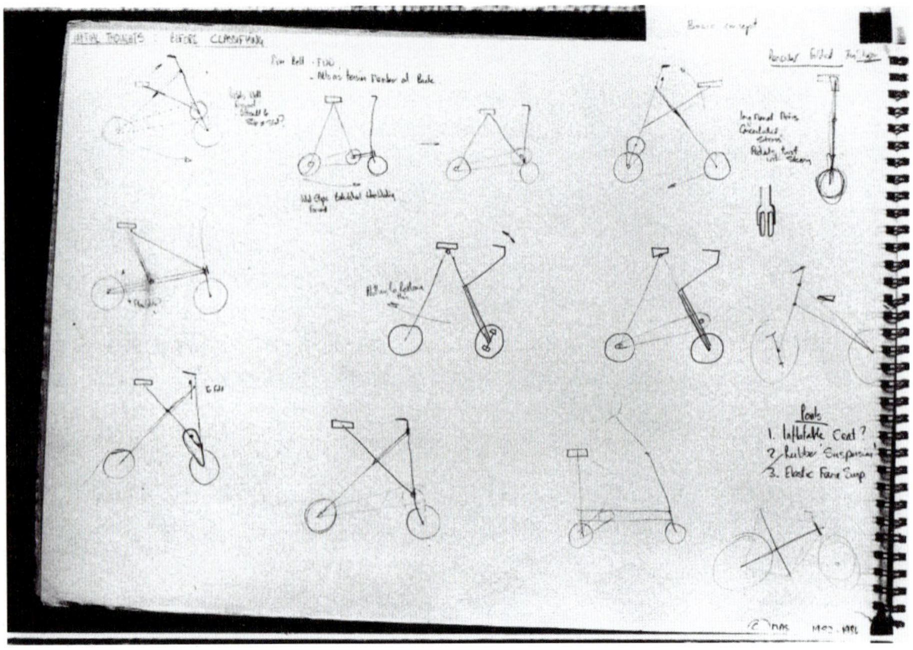

Figure 51 Some initial concept designs for the Strida

This is a page from one of the eight sketchbooks used by Mark Sanders when developing the original Strida folding bike.

Figure 52 Page from Sanders' sketchbooks showing concepts involving adaptation of a car seat belt joint for the bicycle bottom joint

In this type of joint a prong fixed to the car engages a sprung catch on the belt.

Figure 53 The Strida 3 bicycle launched in 2003

Aluminium components are used extensively to save weight and the single gear drive is via an oil-free toothed belt.

Either now or before you start Section 3.2, read the information on Mark Sanders in the DVD notes in the *Block 3 Supplement* and view the DVD sequence 'Mark Sanders: designer and engineer'. This sequence is in two main parts. The first follows Sanders' thinking and designing as he created the original Strida folding bicycle. The second part examines how, after the Strida had been put into production, Sanders improved the design through the Strida 2 and 3 versions.

Once you have viewed the sequence listen to the associated Course Team comments on the DVD and attempt the self-assessment questions below.

SAQ 9

What was the 'problem frame' underlying the basic idea that led to the Strida concept design?

SAQ 10

What methods did Sanders use to translate his basic idea into concept designs?

SAQ 11

What was the source of the idea for the design of the bottom joint?

SAQ 12

Why is sketching so important in Sanders' approach to creative design?

3.1.3 The role of technical analysis

Designing a product such as a folding bicycle is, of course, not just a matter of creativity. It involves choices and constraints, concerning mechanics and materials that are more a matter of *technical analysis*. For example, how did Ritchie and Sanders design the structure of their bicycles to be sufficiently strong and stiff without being too heavy? Was Ritchie correct in choosing steel rather than aluminium for the main frame of his Brompton bicycle and why did Sanders choose aluminium for the Strida?

The *ease and cost of manufacturing* any proposed design is also crucial. Using standard components, such as wheels and gears, is one way to reduce such costs.

Topics such as these will be covered in Blocks 4 and 5 of this course, so I will only touch upon them here. The text below outlines some of the technical factors involved in designing cycle frames. It is included to give you a feel for what is involved without going into the mathematics and engineering.

Technical factors in cycle frame design

A cycle frame is a structure comprising members joined together in a geometrical *arrangement*. Each member is made of particular *material(s)* formed into a geometrical *shape*. The shape, materials and arrangement of its members are the three factors which can be varied in the frame's design. For example, the Strida's main frame comprises *circular section tubes* made of *aluminium* joined into a *triangle* (Figure 53). The Brompton's main frame is a *circular section tube*, made of *steel* in the form of a *beam* (onto which the fork, handlebars, and seat pillar are attached, Figure 49(b)).

A cycle frame should ideally be as *light* as possible consistent with it being *strong* enough to withstand the greatest loads in use without permanently bending or breaking and *stiff* enough to prevent too much elastic (impermanent) flexing or distortion of the frame under load. The loads act both in the plane of the frame and, via the pedals, at right angles to it.

Figure 54 Ordinary (penny farthing) bicycle frame is a beam

Because of the interacting factors, the design of cycle frames is quite complex, so I'll make only a few points here.

The frame of a 19th century Ordinary bicycle consisted of a solid iron backbone with wheel forks at each end forming what is essentially a simple *beam* (Figure 54). To be sufficiently strong and stiff the frame was heavy.

The problem is how can such a cycle frame be lightweight and yet remain sufficiently strong and stiff? There are several possible approaches, as follows.

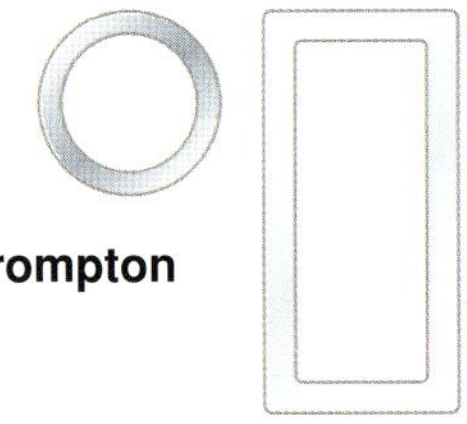

(a) Brompton

(b) Bickerton

Figure 55 Cross-section of the main frame members of the Brompton and Bickerton bicycles

1 Change the geometrical *shape* of the frame members. A frame made from hollow members can be lighter than one made of solid members for the same strength and stiffness. Think of the greater difficulty of bending a piece of tubing compared to that of bending a solid rod made from the same amount of material. Thin-walled tubes make highly efficient structural members. Tubing can vary in diameter and wall thickness. In theory a large diameter tube with thin walls is stiffer than a tube of the same weight of material but of small diameter and thick walls. But there are limits, as a too thin-walled tube can dent and buckle.

   Like most bicycles the frame of the Brompton is made from tubes. In this case the main frame member is a fairly large diameter thin walled steel tube (Figure 55(a)). Tubing is not the only option. If you look at Figure 47 you'll see that the main frame of the Bickerton is a hollow box-shaped member (Figure 55(b)). This box shape was chosen to provide the necessary stiffness because the Bickerton is made from aluminium, which is not as stiff as steel.

2 Choose *materials* that are stronger and/or stiffer for a given weight. Early cycle frames were made of wrought iron, which were replaced by stronger carbon steel frames. In the 1930s frames made from even stronger alloy steel were introduced for lightweight sports bicycles. In the 1990s aluminium alloys, less strong and stiff but much lighter than steel, became the preferred

Figure 56 Mountain bike with frame made from large diameter aluminium tubes

Figure 57 Mike Burrows' third prototype aerodynamic bicycle

This has a monocoque frame moulded from carbon fibre reinforced plastics and front wheel supported on a single fork blade. A rider on the LotusSport, a wind-tunnel-developed version of this design, won the 4000 m pursuit event at the 1992 Olympics.

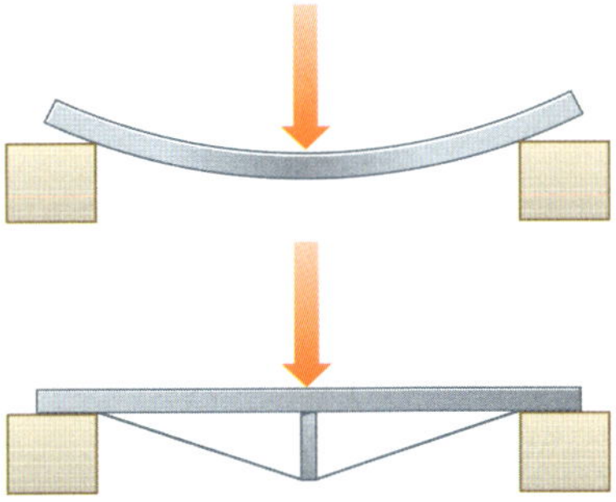

Figure 58 Triangulated structure

material for cycle frames. To provide the necessary stiffness aluminium frames usually have larger diameter tubes than steel ones (Figure 56). When optimised for diameter, aluminium tubes are over three times stronger and seven times stiffer than alloy steel tubing of the same weight. Frames made of exotic materials that are very light as well as being strong and stiff, such as carbon fibre reinforced plastics composites (Figure 57), have been produced in limited quantities.

The Brompton is made from steel because steel could provide the required strength and stiffness of the main frame members using relatively small diameter tubing. Large diameter frame tubes could constrain the design of a bicycle that folds up very small. Steel also offers advantages in resistance to damage and fatigue failure in use. Most important, steel frames can be manufactured using simple welding equipment affordable to a small UK firm.

The Strida's frame is made from aluminium to take advantage of aluminium's light weight (and corrosion resistance). The difficulty of welding aluminium was avoided by using adhesives to bond the frame. Later welded aluminium versions took advantage of the advanced aluminium welding technology available in Taiwan. The triangulated design (Figure 53) of the frame allowed relatively small tube sizes as its reduces the stresses acting on the frame (as explained next).

3 Design the structural *arrangement* so that its members are as far as possible subject to tension (pulling) or compression (squashing) rather than bending or twisting stresses. You can appreciate why it is important to avoid the latter stresses if you consider how easy it is to break a matchstick by bending or twisting it, but very difficult to pull it apart or to compress it.

A structure in which the members are arranged in triangles is subject only to tension and compression when loads are applied in the plane of the structure. For example, a simple beam bridge can be made very much stronger for only a small increase in weight merely by adding a vertical strut and two ties, thus making it into a simple *triangulated* truss (Figure 58).

The diamond frame is a triangle with the front fork attached and two rear triangles supporting the rear wheel (Figure 59(a)). It is therefore very strong and stiff in its own plane.

However, a bicycle frame has to be narrow to fit between the rider's legs. Thus all bicycle frames (diamond-shaped as well as the Brompton and Strida configurations) lack triangulation in depth (i.e. in the plane perpendicular to the paper in Figure 59(a)). The *out-of-plane* forces exerted by the rider pushing on the pedals and pulling on the handlebars can produce severe twisting and bending stresses in the frame. These flex it, as shown in Figure 59(b), wasting some of the rider's energy. Bicycle design is constrained by the need for frames stiff enough to resist twisting and bending stresses without undue flexing. For example, the Brompton's main beam and handlebars were designed not to flex too much when a heavy rider is going uphill and hauling on the handlebars. To achieve this involves a penalty in weight.

Hence the need to design frames that balance light weight, sufficient strength (e.g. to survive a crash) and stiffness, to support uphill pedalling, through appropriate choices of materials and geometry.

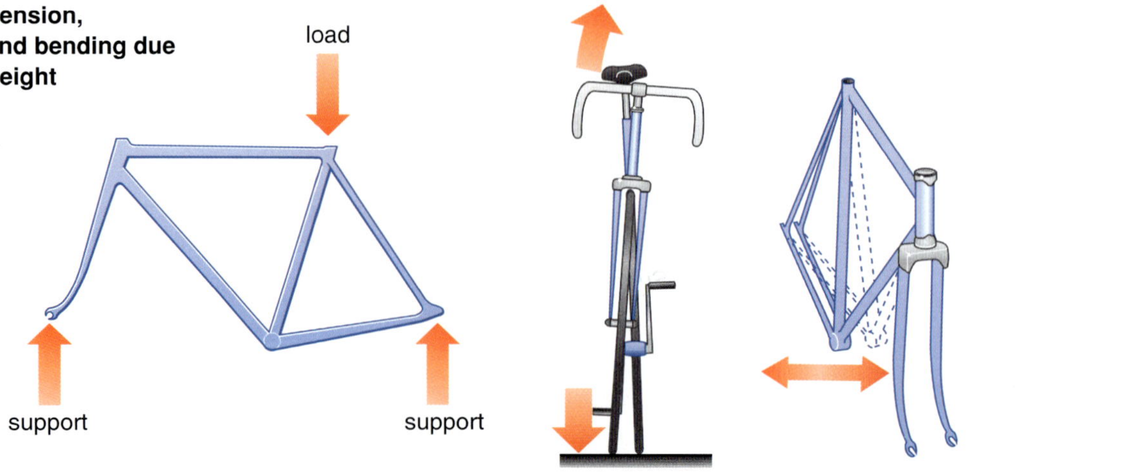

(a) side view: tension, compression and bending due to the rider's weight

(b) front view: twisting and bending due to rider's forces on the pedals and handlebars

Figure 59 Forces acting on a classic diamond frame

3.2 Principles of creative designing

So far you've met a lot of new terms, theories and examples to do with creative designing. I'm now going to ask you in Exercise 8 to look back over Sections 2 and 3 to see if you can distil some creative designing principles from four of the case studies. By doing this you'll see how well you understand the material you've studied so far.

This exercise requires at least 30 to 45 minutes concentrated thought, so you may wish to take a 'breather' before attempting it. As part of your breather and to prepare yourself for the exercise, you could view the DVD sequences on the Cyclone cleaner (James Dyson) and the Strida (Mark Sanders) again. Alternatively, you may wish to postpone part or all of the exercise until you've studied more of the block, perhaps even leaving it until the end. But don't ignore the exercise, it will help prepare you for assignment or exam questions on the 'theory' aspects of this block.

Exercise 8 Elements of creative designing

Below are tables of the elements explicitly or implicitly identified in my four case studies of creative design. Considering one table at a time, try to identify which of these elements were (or were probably) present in each case study.

The most relevant case study sections to look at are:

Section 3.1.1 *Moulton*: From 'Exercise 6 Lessons of the Moulton bicycle' to the end of Section 3.1.1.

Section 3.1.2 *Brompton*: The whole case study.

Section 3.1.2 *Strida*: From the study note about the DVD (*Mark Sanders*) up to the SAQs that follow, plus the related DVD notes in the *Block 3 Supplement*.

Section 2.3 *Cyclone*: From the study note about the DVD (*James Dyson*) up to the SAQs that follow, plus the related DVD notes in the *Block 3 Supplement*.

Spend about 5 to 10 minutes on each table. When you have attempted to complete each table, look at my responses and the discussions that follow. But don't spend more than about an hour on the exercise.

(a) The creative individual

Which of the creative elements listed in the first column of the table were (or probably were) present in each of the case studies listed below? Enter 'Yes', 'No', 'Probably', 'Probably not' or 'Don't know' in the table.

	Moulton (Moulton)	Brompton (Ritchie)	Strida (Sanders)	Cyclone cleaner (Dyson)
constructive discontent				
motivation and perseverance				
expert knowledge				
practical skills				

(b) The creative process

	Moulton (Moulton)	Brompton (Ritchie)	Strida (Sanders)	Cyclone cleaner (Dyson)
problem identification				
preparation				
incubation				
illumination				
verification				

(c) Creative designing steps

	Moulton (Moulton)	Brompton (Ritchie)	Strida (Sanders)	Cyclone cleaner (Dyson)
problem identification				
'problem frame' (my view of the problem is…)				
'primary generator'				
basic idea				
principle proving model				
concept design(s)				
concept selection				
prototype development				

(d) Creative designing methods

	Moulton (Moulton)	Brompton (Ritchie)	Strida (Sanders)	Cyclone cleaner (Dyson)
heuristic approach				
systematic approach				
brief/specification				
immersion				
2D modelling – sketching/drawing				
3D modelling – mock-up(s), prototype(s)				
iteration				
technical analysis				

(e) Associative thinking processes

	Moulton (Moulton)	Brompton (Ritchie)	Strida (Sanders)	Cyclone cleaner (Dyson)
adaption				
transfer				
combination				
analogy				

My response

(a) The creative individual

	Moulton (Moulton)	Brompton (Ritchie)	Strida (Sanders)	Cyclone cleaner (Dyson)
constructive discontent	YES	YES	YES	YES
motivation and perseverance	YES	YES	YES	YES
expert knowledge	YES	YES + learned as needed	YES + learned as needed	YES + learned as needed
practical skills	YES	YES	YES	YES

These creative elements appear to be present in virtually every case (especially where I have entered YES). The only exception is expert knowledge of the problem concerned. But if it was not present at the start, the necessary expertise was learned during the months or years of design and development and was part of the motivation and perseverance required.

b) The creative process

	Moulton (Moulton)	Brompton (Ritchie)	Strida (Sanders)	Cyclone cleaner (Dyson)
problem identification	YES	YES	Yes	An opportunity
preparation	Yes	Yes	YES	Some
incubation	Probably	Probably	Probably	Don't know
illumination	Probably not	Yes	Probably not	No
verification	YES	Yes	Yes	YES

Identification (of a problem and determining to tackle it) and *preparation* (problem exploration and initial attempts at solution) are present in almost all cases. The Cyclone cleaner is slightly different. Dyson spotted an opportunity to improve vacuum cleaners when considering another problem and quickly came up with the basic idea.

Whether *incubation* and a single major *illumination* occurred is less clear. It's likely that each of these designers put the problem aside at times and this should have allowed them to sub-consciously analyse information. But only Ritchie claimed to have had a 'Eureka moment'. These individuals seem to have got their ideas and concepts through many little 'illuminations' while consciously attempting to solve their problem, usually through sketching and physical modelling.

All the designs involved lengthy periods of *verification* (to produce a practical design) once the basic concept had been established. In all cases there was also further improvement to the original design.

(c) Creative designing steps

	Moulton (Moulton)	Brompton (Ritchie)	Strida (Sanders)	Cyclone cleaner (Dyson)
problem identification	YES	YES	Yes	Yes (Opportunity)
'problem frame' (my view of the problem is…)	Yes	Yes	YES	No
'primary generator'	YES	Probably	Yes	No
basic idea	Yes	YES	YES	YES
principle proving model	Probably	Yes	Yes	YES
concept design(s)	Yes	Yes	YES	Yes
concept selection	Yes	Don't know	YES	Yes
prototype development	YES	YES	Yes	YES

Completing this table is not entirely straightforward and your answers may differ from mine, but there are some patterns.

Almost all the inventor/designers had *identified* a problem they wished to tackle, and most had taken considerable time to *explore* it.

In the three bicycle designs this exploration led to a particular view of the problem (the *frame*) plus in two cases a favourite image or objective (the *primary generator*) that gave rise to the *basic idea*. This idea was then embodied in one or more *concept designs*.

The basic idea for the Cyclone cleaner arose more directly, Dyson does not seem to have identified a problem frame or had a primary generator, unless you count always being on the lookout for opportunities to create new and better products. The Cyclone idea arose almost by chance when tackling a different problem by a process of *transfer*. Before embodiment in concept designs this basic idea had to be feasibility tested with a *principle-proving model*.

Almost all the concepts required considerable *iterative* development through a series of *prototypes*.

(d) Creative designing methods

The designers of the Moulton, Brompton and the Cyclone used a mainly *heuristic* approach based on their own experience and methods – essentially repeated *iteration* via design/make/test/modify. The Strida's designer preferred a more *systematic* approach starting with a specification.

Most probably *immersed* themselves in the problem they were trying to solve and used *2D drawing* at some stage, notably Sanders who talked specifically about immersion and made dozens of sketches and drawings to develop his ideas. *3D physical modelling* was universal, notably Dyson with his trial and error testing of thousands of cyclone shapes and experimentation with different hose arrangements, and like Moulton and Ritchie he built several full-size prototypes. Only Moulton and Sanders definitely undertook technical analysis, such as stress calculations. The others may have done so, but it was not central to their designing.

	Moulton (Moulton)	Brompton (Ritchie)	Strida (Sanders)	Cyclone cleaner (Dyson)
heuristic approach	Yes	YES	At times	Yes
systematic approach	At times	No	YES	During development
brief/specification	Probably informal	Probably informal	YES	No
immersion	yes	yes	YES	Yes (at later stage)
2D modelling – sketching/drawing	Yes	Yes	YES	Probably (at later stage)
3D modelling – mock-up(s), prototype(s)	YES	YES	Yes	YES
iteration	YES	YES	Yes	YES
technical analysis	YES	Probably	Yes	No (only at later stage)

(e) Associative thinking processes

	Moulton (Moulton)	Brompton (Ritchie)	Strida (Sanders)	Cyclone cleaner (Dyson)
adaption	Yes	Yes	Yes	No
transfer	Yes	No	Yes	YES
combination	No	No	No	No
analogy	No	No	No	No

Adaptation and/or *transfer* seem to be involved in all of the cases, suggesting that these may be the commonest forms of associative thinking. Although *combination* and *analogy* don't seem to feature, there are other examples in the block where these types of association occurred.

Before you read on, look over this section and in your workfile make your own summary of key points before reading mine.

Key points of Section 3

Section 3 reinforces many of the key points of Section 2. In addition it shows that:

- Individuals who produce worthwhile creative designs usually require great *motivation* and *perseverance*. They usually already possess, or quickly learn, the expert *knowledge* and *practical skills* relevant to the problem.
- In designing, the creative processes of *identification*, *preparation*, *incubation* and *verification* are usually seen. But a major *illumination* ('Eureka moment') may not happen. Instead ideas and concepts typically arise through many minor 'illuminations' often during conscious attempts to solve the problem.
- Effective preparation usually involves *'immersion'* – for example by gathering information, reviewing existing solutions and using similar products, plus constantly thinking about the problem and possible solutions.
- Creative designs can be produced through either *heuristic* or *systematic* approaches. The most common heuristic is that of *iterative* design through repeated cycles of design/make/test/modify.
- *2D sketching/drawing* and (especially) *3D physical modelling* are essential activities in creative designing. *Technical analysis* may also be involved, but for less complex product design problems may be done informally from experience or not be fully carried out.
- Creative concepts involve making a *compromise* between (or ideally fully *resolving*) the often *conflicting requirements* of the design.

(Learning outcomes 1.1–1.4, 2.1, 2.2 and 3.1)

4 Creative designing – systematic and team approaches

In the previous sections I looked at the ways that creative individuals designed innovative products mostly using their own heuristic approaches.

In this section I shall look in more detail at *systematic* approaches to concept design, mainly for product and engineering projects. Systematic design approaches, some argue, produce better concepts because they encourage the designer to divergently explore the solution space. For example, in his book *Total Design* Stuart Pugh writes:

> A tendency to 'cut and run' is as common with students of design as it is with professionals. ... you need as many ideas as you can possibly generate [...] remember that ... the wrong choice of concept ... can rarely ... be recouped by brilliant detail design.
>
> (Pugh, 1991, pp. 69, 73)

You may recall Pugh's example of the systematic design of a car horn in Section 1.4. In that many concepts based on different technical principles were identified and the best selected for development using the criteria in the specification.

Systematic methods are sometimes employed by individual designers, but by providing a framework for the design process are most useful to those who are *learning to design* and to designers working in *teams*. Mark Sanders, for example, employed a systematic approach (when as a student) he designed his folding bicycle. Again I'm going to use examples of bicycle design, but this time focusing on team design of components and accessories.

4.1 Mountain bike 'Splashguard'

The example below is based on an account (from Ullman, 1992, pp. 104–157) of a design project undertaken by a team of engineering students. Professional designers probably would not have tackled such a relatively simple project so thoroughly – the example illustrates elements of a systematic approach to concept design as attempted by learner designers, from which you can learn.

4.1.1 Design brief

Mountain bikes have become the most popular design of bicycle in Western markets. They have strong and stiff frames and thick tyres, and often wheel suspension, designed for riding off-road including on mountain tracks (see Figure 56). In practice they are more often used on roads for commuting and leisure. Most mountain bikes lack mudguards, and the brief given to the team identified the need for a low-cost, detatchable device to protect the rider from water and dirt thrown up by the rear wheel.

4.1.2 Customer requirements

The design team first identified the customer requirements for the 'Splashguard' (see Box 2) using methods similar to those discussed in Block 2 (e.g. talking to users, specifying benefits and requirements).

Box 2 Shortlist of Splashguard customer requirements

Functional performance

- Keeps water and dirt off rider.
- Quick and easy to attach and detach, including when bike is dirty.
- Not retain water/mud/debris.
- Not damage bicycle.

Structural integrity and safety

- Not rattle, wobble or bend.
- Lightweight and durable.
- Won't release accidentally.
- Mustn't interfere with bicycle operation even if it fails.

Spatial constraints and appearance

- Fits most bicycles.
- Not interfere with rider or any bike components.
- Contemporary look, fashionable colours.

Time and cost

- Developed in 3 months, market in 12 months.
- Manufacturing cost under £2 per unit.

Manufacturing

- Manufactured using existing facilities.
- Use standard components where possible.

As noted in Block 2 Section 10, some of the qualitative customer requirements in Box 2 would normally be translated into a quantified *performance specification* as part of the product design specification. For example, the requirement 'Quick and easy to attach and detach' might after some user research be quantified as 'Can be attached and detached by 95% of users in less than 1 minute'.

4.1.3 Function analysis

The basis of many systematic design methods is that concepts should arise from consideration of a product's *functions* (i.e. the essential requirements in its performance specification) and, since there are usually many ways of fulfilling each function, there exist many possible concept designs within the solution space.

A function is what is essential to the practical operation of the device. Function specifies *what* the product must do, whereas the concept design conveys *how* the product will do it.

Exercise 9 Identifying functions

Look at Box 2. What are the essential functions of the Splashguard that must be met by any satisfactory concept design?

Spend 5 minutes making notes in your workfile before looking at my response below.

My response

I would say that the essential functions of the Splashguard based on the customer requirements are: 'An easily detachable device that can keep water and dirt off the rider of a mountain bike without getting clogged or interfering with the bike's operation.'

You might have included other requirements, such as the device being 'lightweight' and 'durable'. Such requirements are not strictly functions, since they are not essential to the practical operation of the device. They will be considered later when evaluating alternative concepts.

Designing the Splashguard was then based on the technique of *functional decomposition* (see for example, Ulrich and Eppinger, 2000), followed by generating concepts from the identified functions.

Functional decomposition is a fairly advanced technique and I won't attempt to teach it here. Essentially, it identifies the functions of the independent subsystems that make up the total product (in terms of each sub-system's role in converting energy, material or information from one form to another). The technique can also be applied informally as an aid to thinking in functional terms.

The team's functional decomposition of the Splashguard indicated that it has two sub-systems, whose functions are:

1. to prevent water and dirt thrown up by the rear wheel from soiling the rider;
2. to allow the device to be quickly and securely attached to the bike until it needs to be detached.

These functions could be provided by many different concepts – from a simple mudguard to a complex water and dirt absorption system.

4.1.4 Concept generation

Concept generation uses the functions identified to foster ideas. There are two steps. The first is to create a wide variety of ideas, or technical principles, that can satisfy the functions of each of the sub-systems identified. The second is to combine these ideas into one or more concepts that meet the functional requirements of the total product or system.

There are many ways of generating ideas to meet functions, including those I identified earlier, such as *adaptation*, use of *analogies*, etc. There are also creative problem solving techniques to aid designers in generating ideas, some of which I shall be introducing in Section 5.

One such technique is *morphological analysis*(which you tried in Section 1, Exercise 5 and will be fully described in Section 5.3.2 and on the DVD). In the Splashguard example, the technique was used to list different ideas that could provide the functions of the device's sub-systems and then to generate different combinations of these into concept designs (Box 3). Generating the ideas listed in Box 3, did not happen automatically as part of the technique, but depended on the design team's creative thinking skills.

Box 3 Some of the design team's ideas to meet two functions of the Splashguard's sub-systems

Function 1. To prevent water and dirt thrown up by the rear wheel from soiling the rider.

Ideas:

- device on rider;
- device between rider and wheel;
- device next to wheel, like mudguard;
- scrape water and dirt from wheel;
- prevent water and dirt from getting to wheel.

Function 2. To securely attach and detach the device.

Ideas: screw; suction cup; spring clip; Velcro; pin; magnet; elastic.

Combining the ideas given in Box 3, and others, using morphology can produce many different concept designs: familiar (e.g. *mudguard* fixed with a *spring clip*), novel (e.g. waterproof sheet *on rider's* back fixed with *Velcro*) and odd (e.g. device powered by the bike to blow the water and mud away). Some evaluation is necessary at this stage to weed out overlapping or unworkable ideas, but without prematurely discarding unusual concepts that may prove useful. The process of combining and evaluating should throw up further ideas.

So far designing the Splashguard has been text-based, mainly comprising lists and charts. Although the concept designs (developed e.g. by morphology) may still be quite abstract, rough 'back-of-the-envelope' sketches now begin to be useful. Often a single-view sketch is satisfactory, but if the drawing needs to show the concept viewed from more than one direction simple perspective or orthographic sketches can be used (see the *Modelling Workbook*, Section 4). Again, the process of sketching should lead to new ideas and concepts.

The Splashguard concepts in Figure 60 are seven of the many generated by the student design team:

Concept 1: a poncho or raincoat.

Concept 2: a device between rider and wheel that attaches to the seat post and/or seat.

Concept 3: a plastic window-shade device, spring-loaded to roll up in a tube when not in use. The tube scrapes off the water/dirt as the Splashguard rolls up inside it.

Concept 4: the 'Sissy Bar'; another device between rider and wheel.

Concept 5: a device to brush water/dirt from the wheel.

Concept 6: a luggage rack with an attached mud flap.

Concept 7: a detachable mudguard (fender).

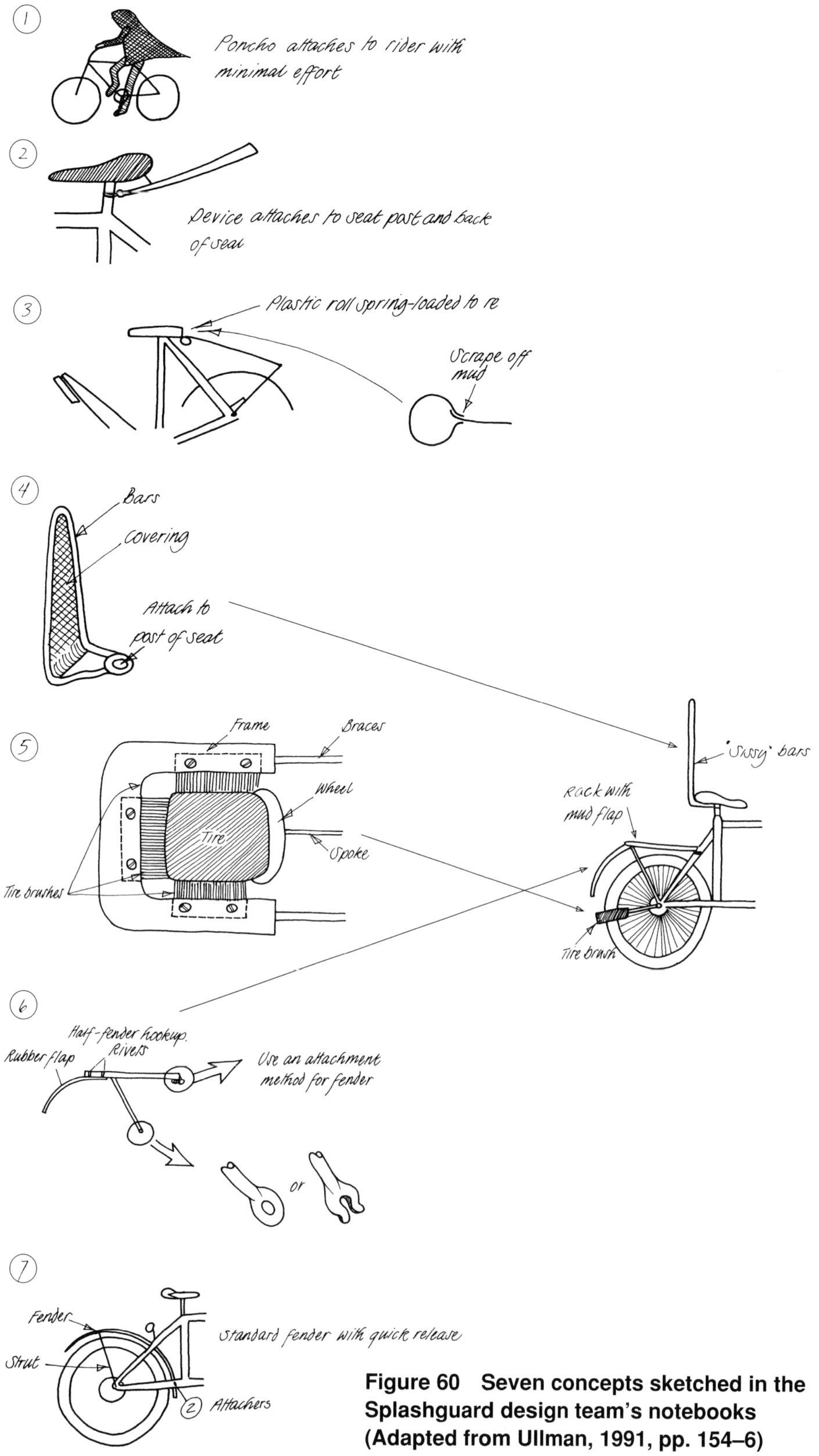

Figure 60 Seven concepts sketched in the Splashguard design team's notebooks (Adapted from Ullman, 1991, pp. 154–6)

Exercise 10 Sketching concept designs

Look at the functions listed in Box 3 and the concepts in Figure 60. Try to produce at least one other idea and make a rough sketch or a diagram of the concept in your workfile.

Spend about 15 minutes on this exercise before looking at the Course Team's response. Your sketch or diagram doesn't need to be any more polished than those in Figure 60. If necessary consult the *Modelling Workbook* (*e.g.* Section 4, Figure 42 or Section 7, Figures 97 and 98) for guidance on rough sketching.

Course Team's response

A member of the Course Team thought of an inflatable barrier between wheel and rider fixed to the saddle or luggage rack that could be inflated with the bicycle's pump (Figure 61).

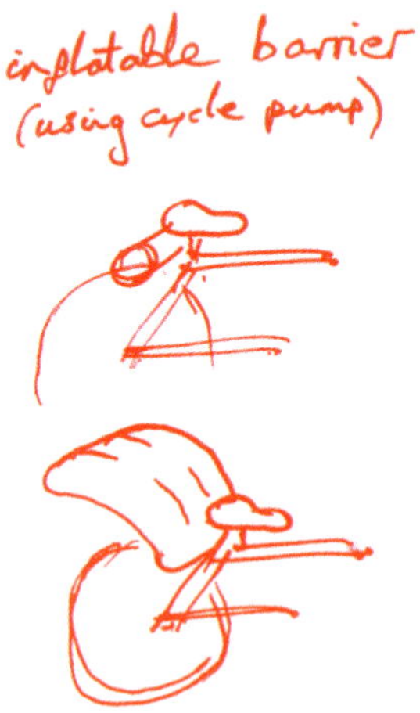

Figure 61 Another concept for the Splashguard (Steve Garner)

4.1.5 Concept selection

Having generated a number of possible concepts, it is necessary to select one or more for further development. After the divergence of concept generation, this is a *convergent* phase of the designing process.

The most useful concept selection methods include:

- *Go/no-go evaluation* by checking if each concept satisfies the customer requirements. For example, Concept 1 the poncho gets a 'no' to the first requirement in Box 2 'keeps water and dirt off rider', as water/dirt could be thrown up under it (Figure 62) – so it's 'no-go'. Concepts 2, 3, 4, 6 and 7 produced 'yes' or 'maybe' answers for all the requirements and so were 'go' for further development and evaluation.

Figure 62 Go/no-go evaluation

Concept 1 (The poncho) could fail the essential customer requirement of keeping water and dirt off the rider – so it's no-go.

- *Technical performance/feasibility testing*. The feasibility of the tyre brush (Concept 5) was uncertain. So the team tested it with a principle-proving model comprising a bicycle with various brushes and sponges applied to the wheel as water was sprayed onto it. None prevented all the water being thrown up, so this concept got a 'no-go'.
- *The selection/evaluation matrix* (e.g. the car horn matrix in Figure 17). This method systematically scores each concept relative to a standard design called the 'datum' (in this case a detatchable mudguard) using the customer requirements list (or performance specification). This gave the best score to Concept 2, which was taken to embodiment design stage.

> The selection/evaluation matrix technique is outlined in Section 5.3.3. When you study that section, there is also an interactive exercise on the DVD that guides you through the technique to evaluate the Splashguard concepts.

4.2 Bicycle backpack carrier

This second case concerns the design of another accessory, a device for mounting and transporting a hiker's backpack on a mountain bicycle.

This problem was given to a team of three designers employed by the consultancy firm IDEO, who agreed to have their thinking and designing recorded for an experimental research study. Nigel Cross analysed the creative process that led to a concept design for this device. Here are some extracts from his analysis:

> The team's approach to the given task was a relatively rational and systematic one. Very early in the session they planned a design strategy ... a variation of a conventional model of the engineering design process. ... (1) exploring the problem and writing a performance specification; (2) generating a range of concepts; (3) evaluating and selecting the most promising concept; (4) developing the concept into a detailed design; and (5) communicating the final proposed design. [These five phases are an example of a *systematic design process* typically used in product and engineering design.]
>
> [...]
>
> The team ... developed the problem into three sub-problem areas: (1) the position of the rack device relative to the bicycle; (2) joining mechanisms between (a) the backpack and the rack and (b) the rack and the bicycle; and (3) materials necessary for making the rack device.
>
> In each case, the team members explored *problems and solutions together* by proposing concepts (sub-solutions) for each sub-problem and evaluating/discussing the implications and possibilities of each concept. For example, their thoughts about the positioning of the rack and its supports on the bicycle frame raised issues regarding riding stability, ergonomics of use, concerns about the weight of a full backpack and user behaviour. In general, they argued from *form to function*, rather than vice versa.
>
> One of the significant issues that arose during discussion concerned the hazard caused by the backpack's shoulder straps if they were allowed to

dangle into the bicycle wheel. After generating random concept lists, the team members reviewed each list to eliminate unsatisfactory concepts and identify preferred ones. As they went through the pack-to-rack list, the 'bag' concept was stressed as a solution for holding all loose straps. It was then that the tray concept suddenly appeared:

[I, J and K are the three design team members.]

> I: Bag: put it in a bag. We're going to need some sort of thing to do something with those straps.
>
> K: To get this out of the way …
>
> J: So it's either a bag, or maybe it's like a little vacuum-formed tray kind of for it to sit in.
>
> I: Yeah, a tray, that's right, ok.
>
> […]
>
> J: Maybe the tray could have plastic snap features in it, so you just like snap your backpack down into it.
>
> K: Snap in these (backpack) rails.
>
> J: It's a multi-function part.
>
> K: You just snap in these rails.
>
> J: Yeah, strap the rails into the tray there.
>
> I: ok
>
> J: It takes care of the rooster-tail problem …

In this segment of the discussion we see the key concept – the tray idea – proposed, accepted and then modified, developed and justified. As well as securely holding the backpack, a tray would solve two particular problems: the dangling straps and the 'rooster-tail' problem – i.e. the water/mud spray thrown up by a mountain bicycle wheel, which … would dirty the backpack unless it were protected. […]

Did the tray idea come from nowhere? … Related concepts mentioned earlier include injection-moulded plastic as a material and flat plastic forms to be used as the rack device … Before the tray idea was first expressed, its originator, Designer J, referred to a similar rack idea: 'It looks like everything we're looking at now is wire-form, but a friend of mine suggested … an injection-moulded rack that would … fold down.' […]

'the tray idea … is recognisably a bridging concept between problem and solution that … resolves a variety of goals and constraints.'

The generation of the tray idea was perhaps aided by the earlier consideration of a more extreme form of containment: a bag. The idea seemed to focus on one particular problem (containing the straps) as most significant, then was quickly elaborated to satisfy a range of other problems and functions. It is recognisably a bridging concept between problem and solution that synthesises and resolves a variety of goals and constraints.

(Cross, 1997, pp. 430–432)

Having described the activities that led to the concept design shown in Figure 63, Cross discusses the thinking by which the creative leap to the idea of a tray to contain the backpack emerged. These include creative processes you met earlier, a *problem frame* (the need to contain dangling straps) and probably either *adaptation* (flat panel into tray), *combination* (panel + bag = tray) or *analogy* (of bag with tray). This inspired a *basic idea* (the tray) that was then developed into the concept design. The difference from the individual designer's thinking process is that in a small team the creative ideas flowed quickly, interacted with and built on each other.

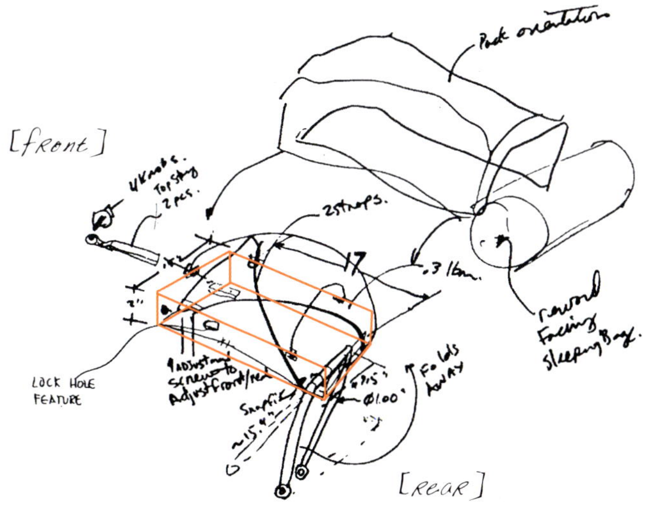

Figure 63 The design team's concept design for a mountain bicycle backpack carrier; the tray is highlighted in orange (adapted from Cross, 1997, p. 428)

This observation of creative designing showed that the designers adopted a systematic process to structure their work. But they didn't systematically follow logical steps from overall function and sub-functions to generate multiple concept designs (as was attempted in the functional decomposition method for the Splashguard). Instead, after the crucial step of analysing the problem into sub-problems (i.e. the rack positioning, fixing and materials, and how the backpack was to be secured to it) the designers informally brainstormed alternative overall concepts and sub-solutions (bag, tray, connectors, etc.). Within two hours they had produced a concept design. When an idea arose that simultaneously bridged all, or most, of the sub-problems and sub-solutions – in this case the tray idea – this provided the creative flash of insight.

This example demonstrates again that the *resolution* of conflicting problems and sub-problems is the essence of creative work. Experienced designers are often able to produce a concept design quite quickly by identifying the key sub-problems and the most promising ideas for resolving them.

SAQ 13

What lessons about creative concept design processes can you identify from the Splashguard and backpack examples in Sections 4.1 and 4.2?

4.3 Creative design teams

Working in a team provides clear advantages in terms of the requirements for creative designing. The necessary specialist knowledge, broad experience, practical skills and hard work can be shared. As the backpack example shows, a team provides the opportunity to discuss, be stimulated by and build on each other's ideas. It's possible to develop ideas through informal discussions or visual brainstorming sessions which team members can then develop individually. Team members can provide the constructive discontent, the motivation and stimuli needed for creative work. They can also provide the critical evaluation that prevents individuals getting too fixed on a single idea.

Although some creative designers work mainly alone, most modern design projects are too big and complex to be undertaken by an individual. Such projects require teams of designers and others, such as managers, marketing and manufacturing specialists, working together. These teams can be small, perhaps two to three designers for a web design project, or medium sized, say ten designers and engineers plus a project manager for a domestic appliance project such as the Dyson DC01 Cyclone cleaner shown on the DVD. Some teams may involve hundreds, or even thousands, of people, e.g. for the design of a new model of motor car or aircraft.

But whether large or small, the nature of the *team* and the *organisation* within which it works affects its ability to produce creative designs. Key aspects of the team and the wider organisation include:

- *Team composition*. An effective team includes people with specialist knowledge and those with team skills in managing and interacting with other team members.
  To provide *specialist knowledge* in different elements of a problem, a design team is normally multi-disciplinary. The different team skills are provided, for example, by member(s) who are good at gathering and analysing information; by members who can generate ideas; a project manager who allocates tasks and enables decisions; and a member to provide finance and project support. Except in large teams, each member will usually have both specialist knowledge and one or more team roles.
- *Communication between team members*. A key element of team creativity is communication and cross-fertilisation of ideas. To foster informal and formal communications, design centres have been established by firms such as Philips to house team members physically close to each other. But increasingly, design teams may be located in different parts of the world and communicate electronically by email, shared CAD systems, Powerpoint presentations, etc.
- *Organisational culture and creative climate*. It's important for a team to work in a 'climate' in which creativity can flourish. In creative organisations managers will support new ideas, unofficial project groups may be tolerated, and there may be time allowed or reward systems for innovation. In firms doing creative work, dress codes may be informal and the working environment may provide for relaxed communication (see Figure 64).

Figure 64 Brainstorming session at IDEO's California studio

Wild ideas, building on the ideas of others are encouraged and recorded on Post-It notes and flip charts.

- *Organisational structure*. Co-ordination of the contribution of different individuals and departments in an organisation is essential to a creative team. Products designed in organisations where there is division and poor co-ordination between marketing, design and manufacturing departments are often late and require redesign. To avoid such problems, design projects are usually undertaken either by a dedicated *project team* temporarily assembled for the task, or in a *matrix* structure that can provide the benefits of a project team with access to the specialist departments in an organisation.

I shall not be considering these managerial aspects of creative designing in any more detail. However, you can get a flavour of how a creative team at IDEO, a large international research, design and development firm, is organised and managed from the DVD sequence 'IDEO: creative design teams'. You already learned in Block 2 about IDEO's user-centred design approach in creating Amtrack's new train service and the BBC's digital radio service. Examples of IDEO's projects shown on the DVD include the first laptop computer and a player for music downloaded from the Internet (Figure 65).

Figure 65 IDEO's Audible Mobile Player

This is designed to transfer and play material from the Internet; the design is based on the metaphor of the ear.

Before starting the next section look at the DVD notes about IDEO in the *Block 3 Supplement* and then view the DVD sequence 'IDEO: creative design teams' and listen to the associated Course Team comments.

The 'Philips: creative product design' DVD sequence is also very relevant. It shows the use of visual brainstorming and associative thinking by a team at Philips Design when creating a new product line. After viewing the sequence on IDEO, look at the DVD notes on Philips in the *Block 3 Supplement* and (re)view the sequence 'Philips: creative product design' and listen to the associated Course Team comments. Then attempt the following SAQ.

SAQ 14

Identify the common factors in the management and design practices of the teams at IDEO and Philips Design shown on the DVD.

Key points of Section 4

Systematic design methods encourage the designer or team to divergently explore the solution space to find an optimum concept.

Systematic methods (for engineering and product design) introduced in this section typically consist of several steps:

1. The problem is explored, customer requirements are identified and a performance specification drawn up.
2. The essential functions of the product and its sub-systems, or key sub-problems, are identified.
3. The identified functions, or sub-problems, are then used as the basis for generating many solution ideas.
4. Concept designs are developed through sketching and modelling.
5. Concept(s) are selected for development by evaluation against the specification requirements.

Systematic methods are most useful to those who are *learning to design* and, when applied *informally*, to experienced designers working in teams. A systematic approach may be required in large organisations to ensure project management.

Systematic approaches used informally by experienced design teams involve designers identifying and exploring key sub-problems and sub-solutions together in an attempt to find a concept that resolves the sub-problems. In a team the creative ideas can cross-fertilise and build on each other to rapidly converge on a solution.

Designing in a team has many advantages for creativity. The knowledge, experience, skills and hard work can be shared. Team members can provide the motivation and external stimuli needed for creative work and prevent individuals from getting too fixed on a single idea.

The design team, and the organisation in which it works, have a crucial effect on creativity. Creative teams tend to be multi-disciplinary with members that offer specialised knowledge and/or team skills. The teams are organised to facilitate communication and to co-ordinate departmental contributions; and work in an environment that supports creativity.

(Learning outcomes 1.3, 1.5, and 2.2)

Part 2
Practice

5 Practical creativity for design

So far you've learned about theories of creative designing. In this section we get down to practice! You'll see how what you've learned can help enhance your creative ability, for example when attempting the TMA for this block.

Before studying this section, I suggest that you quickly look through the TMA and associated TMA support materials in the *Block 3 Supplement*.

5.1 Creative designing requirements

We'll start with a revision exercise. It should help you summarise everything you've learned so far about the requirements for creativity in concept design and so help you understand and apply the practical approaches and techniques in this section that stem from these requirements.

Exercise 11 Creative design requirements

On the basis of what you've learned so far, list the key requirements for an individual or team to produce creative work, with particular reference to concept design.

Spend 10–15 minutes on this, referring to the key points of Sections 2–4 (and, if you have attempted it by now, Exercise 8 in Section 3) before looking at my response below.

My response

Key requirements for creative work are:

1. *Constructive discontent.* Not being satisfied with what exists, a desire to improve things or produce something new.
2. *Motivation.* Interest and persistence in tackling a problem or task. A willingness to work hard, often in the face of difficulties.
3. *Knowledge and experience*
   - (a) *Deep*: as much knowledge and experience as possible in the field of the problem or task concerned and the ability to quickly acquire relevant new information. The more specialist knowledge and skill the individual or team has, the richer is the 'raw material' available for creative work.
   - (b) *Wide:* knowledge and interest in many fields both related and unrelated to the problem area, to provide new associations and combinations.
4. *Preparation.* 'Immersion', problem exploration or task clarification to discover as much as possible about the problem or task and possible solutions.
5. *(Re)defining the problem.* The ability to analyse a problem or task and identify a productive starting point: a '*problem frame*' (e.g. an essential

function or fundamental issue to be tackled) and/or a '*primary generator*' (an idea, image or objective that generates an initial solution to a problem, or the initial solution itself.)

6 *Associative thinking*. Open to using and being inspired by ideas, objects and solutions from the same and different fields.

7 *Divergent/convergent thinking*. An ability to think both divergently (to generate ideas and concepts) and convergently (to select good solutions) and to switch between the two modes.

8 *Resolving conflicting requirements*. The ability to resolve conflicting requirements into an optimum or elegant solution.

More specific requirements for creative (concept) designing include:

9 *2D and 3D modelling.*

(a) The ability and willingness to sketch and draw, at least roughly – visual thinking.

(b) The practical skills to make simple mock-ups, models and prototypes – 'thinking with the hands'.

Finally, specific requirements for creative design teams include:

10 *Team composition, structure and environment*. The members should have relevant knowledge and team skills; communicate freely; be organised to co-ordinate members' or departmental contributions; and work in an environment that supports creativity.

The above requirements for creativity probably seem rather daunting. But don't forget we've been looking mainly at highly innovative designs and inventions. Most designing leads to much less innovative 'variant' designs and so the creative requirements are less severe. Also working in a team can dramatically reduce the knowledge and skills each individual needs.

As you've seen, experienced designers and design teams usually rely on their knowledge and experience to find a starting point and to converge on a few worthwhile ideas and concepts within the solution space. They use heuristic approaches, such as iteration, modelling and discussion, to evolve draft concept designs. Learner designers, however, may also need the aid of systematic design methods and creative problem solving techniques that formalise the heuristic approaches and ensure a wider exploration of the solution space before converging on an idea or concept.

> Learner designers – probably including many of you on this course – when attempting to create concept designs should start by applying the approaches used by *experienced* designers outlined below in Section 5.2. (If you are an experienced designer or engineer, you can also learn from the approaches outlined in Section 5.2.)
>
> If you are not satisfied with the results of applying the experienced designers' approaches in Section 5.2, try using the framework provided by *systematic design methods* and the aid of *creative problem solving techniques* to explore problems and to generate and select concept designs. A few such methods and techniques are outlined in Section 5.3 below and taught via interactive exercises on the DVD.

5.2 Experienced designers' approaches to creativity

5.2.1 Constructive discontent and motivation

To provide constructive discontent it is important to be *sensitive to the strengths and weaknesses of existing products and systems*, for example by observing and thinking how they could be improved. The user research methods in Block 2, Section 6 are useful here.

Motivation is generated by working on problems or tasks *in which you are interested*, often an area in which you or someone you know has experienced a difficulty or hazard. Motivation might also be generated by *being set a task that you want to perform well* – perhaps the TMA for this Block! Whichever the starting point, motivation is affected by your assessment of the likelihood of succeeding at the task, i.e. *tackling problems that you (or your team) consider are within your capacity*.

5.2.2 Knowledge and preparation

Knowledge is crucial to creativity. The knowledge needed to tackle a problem is more likely to be available if you work in areas in which you have *at least some existing knowledge and experience* or in which you can quickly learn or gather the necessary information.

As you've seen, creative ideas sometimes arise through the process of *preparation, incubation, illumination* and *verification*. Provided you are well enough prepared with different aspects of a problem, the idea for a solution may occur spontaneously due to subconscious patterning of information in the brain.

Whether or not it leads to a breakthrough 'illumination', preparation is essential for creativity. Proper preparation requires exploring and clarifying the problem to be solved, gathering as much information about it as possible, lots of hard thinking, and often making several unsuccessful attempts at a solution. On the DVD sequence which you viewed in Section 3, Mark Sanders describes this as 'immersing yourself in the problem'. The idea is to saturate the mind with the problem and possible routes to a solution.

How is this to be done?

Gathering information

Sometimes an experienced designer will know enough to be able to solve a problem without any further information other than that provided in the brief. However, except perhaps in routine design tasks, it will almost always be necessary to gather further information.

It's impossible to know in advance what information will prove relevant to a particular problem. You've seen that creative ideas often come from putting together information from previously unrelated fields. So in theory anything could be relevant. But of course it isn't possible to find out everything there is to know to solve a single problem. So where do you start?

State of the art review

Often a designer will start by reviewing the 'state of the art'. This means examining existing solutions in the same or similar areas, by looking through manufacturers' catalogues, visiting shops, showrooms or museums, reviewing patents, reading appropriate books, magazines, trade journals, and so on.

Each designer will have their preferred sources of information and inspiration. Mark Sanders explains that, as part of 'immersing himself', he looked at as many publications showing folding bicycle designs as he could find. What designers are doing is to expose themselves to many ideas and technical principles that might provide them with a solution.

Internet searching

The Internet provides almost instant access to a vast source of information on virtually any topic. But it's essential to hunt for what you want using a good search engine. My current favourite is Google **(http://www.google.com)**, which rapidly searches *billions* of Web pages. With the Internet there's always the danger of being swamped with irrelevant or unreliable information. To avoid this you should learn how to conduct Internet searches.

> The Open University Library provides very useful guidance on how to search for information on the Internet. See the *Block 3 Supplement* for details of how to access this guidance and for recommendations on which search engines to use.

Consulting people and organisations

Consulting other people and organisations is another very valuable thing to do. Most likely these will be experts or other designers in the field. But, as Block 2 showed, users, sales-people, retailers or suppliers can also provide very relevant ideas and information.

Other ways of preparation

Observing existing products in use and becoming a user oneself (applying the 'user trips' technique presented in Block 2, Section 6) are also good ways of immersing oneself in a problem as well as being valuable sources of ideas and information.

But don't forget that just collecting information in the hope that a solution will occur 'spontaneously' is not enough. It is necessary also to make *deliberate attempts to create solutions.* Even if these are unsuccessful, they are an essential part of preparing the mind to generate the 'real' solution.

Exercise 12 Information sources

Consider the problem of redesigning a vacuum cleaner to make it more energy efficient, while maintaining its performance and usability. What sources might you use to inform yourself about the problem and provide ideas for a solution?

Spend 10 minutes making a list of information sources in your workfile before looking at my response below.

My response

(This is the first of six checklists that should be useful for the Block 3 TMA)

Here is my checklist of information sources for a problem such as improving the energy efficiency of a vacuum cleaner:

Checklist	Information sources
1	Internet search (concerning the principles of, and ideas for improving, the energy efficiency of domestic products; and the history, design and technology of the product and of devices or systems with a similar function).
2	Consult librarians, search/browse libraries, bookshops and the Internet for relevant patents, books, and magazines.
3	Contact firms and search the Internet for manufacturers' catalogues and technical information.
4	Visit shops and museums (where the product, or other energy efficient domestic products, are displayed).
5	Talk to users (of the product and similar devices or systems). Use the product, or similar devices or systems.
6	Contact relevant trade bodies, experts and academics.

Thorough preparation, e.g. by gathering information, is vital for most designers, especially those with limited knowledge and experience.

5.2.3 (Re)defining the problem

Defining the right problem to solve is crucial to coming up with a creative solution. This sub-section looks at three approaches you've met:

- framing the problem;
- essential function/performance specification;
- the primary generator.

Framing the problem

You've seen that the ideas and concepts produced by the bicycle designers considered earlier depended on their particular viewpoint or definition of the problem, the so-called 'problem frame'. For example, Andrew Ritchie's problem frame to achieve an ultra-compact folding bicycle was 'to bring the handlebars, seat, front and rear wheels to one spot'. However, arriving at a problem frame usually involves thorough immersion in the problem (as discussed in Section 5.2.2). It may have required redefining the problem as initially perceived, or as given in a brief.

> For both learners and experienced designers, preparation by 'immersion' is a good method of 'framing' a design problem.

Essential function/performance specification

Block 2 showed the importance of establishing a specification at the start of designing.

On the DVD you saw how Mark Sanders drew up a specification before designing the Strida folding bicycle and in Section 4.1 you read how the Splashguard project team identified the product's essential functions at the beginning of designing.

Another example of thinking in terms of performance or function is Harry Bickerton's approach when designing his folding bicycle (Figure 47), which you may recall was the inspiration for the Brompton.

Bickerton did not wish to be 'hidebound by the bicycle' and at the concept stage was prepared 'to consider any promising solution to the problem'. He therefore started by writing a *performance specification* that set out the requirements of something whose essential function was to be 'a new form of portable personal transport'.

Bickerton's ten-point specification (quoted in Roy, 1983, p. 110) included the following requirements:

1 The vehicle must be as light and compact as possible … and no more difficult to carry … than a medium sized suitcase.

2 It must enable its user to cover distances up to 5 miles (8 km) … at least four times faster than walking.

3 It must not require standards of physical ability and skill above normal.

[…]

8 It must be suitable for quantity production and sale at a price … not more than twice that of a good quality bicycle.

Bickerton considered alternative concepts that might satisfy his specification, including large wheel roller skates and even a personal airship. He made sketches and some experimental prototypes and considered the pros and cons of each alternative against the specification. In the end he concluded that a folding bicycle best met the specification, but was satisfied he had fully explored the solution space.

Arriving at the essential function(s)/performance specification of a proposed design isn't always easy. A way of approaching the task is discussed in Block 2 Section 10, using examples of a still camera, an electronic screwdriver and a portable fax machine, and in this block for the example of the Splashguard in Section 4.1.

Exercise 13 Thinking in terms of function/performance

Return to the problem of designing an energy efficient vacuum cleaner. How might you redefine the problem in functional/performance terms in order to open up the solution space?

Spend a few minutes thinking about this and making notes in your workfile before looking at my response below.

My response

Checklist	Function/performance
1	What is the essential function?
	The essential *function* of a vacuum cleaner is to collect dust and dirt from floors and carpets. You are now not constrained to think just of vacuum cleaners, but the solution space opens to include any device, machine or system that can collect dust and dirt using less energy, such as an improved carpet sweeper.
2	What are the performance requirements?
	Other crucial functions and requirements (e.g. cleaning upholstery, lack of noise) may need to be identified by using the methods in Block 2, Section 10 for compiling marketing and performance specifications.
3	What are the essential sub-functions?
	The functional approach often works better at the *sub-system* or *component level.* For example, the function of the electric motor on the vacuum cleaner is to provide power to suck up dust. Then you are free to consider different methods of providing this power using less energy, such as a pneumatic motor, foot pedals, and so on. You could do the same for the function of the dust-bag and the other parts.

For both learner and experienced designers, thinking in terms of function/performance is a useful method of problem definition and concept generation.

The primary generator

You've also seen that architects and designers often start with an initial solution concept – the primary generator. This concept is used to explore and test the realities of the problem in order to understand and define it better. The initial concept can then be modified, or rejected and a new one tried.

The primary generator usually stems from the knowledge, ideas and favourite objects that a designer has accumulated from long experience. For example, an architect or product designer might draw on an image from a favourite film, product, building, or car to create an initial concept design. Recall from Section 2 Jonathon Ive basing his design for the iMac range of computers on the adjustable Anglepoise lamp. An engineer might start from objects or images that embody the latest technological developments.

A solution-focused strategy to problem definition can be used by learner designers, but is probably more suited to those with experience.

5.2.4 Associative thinking

You learned in Section 2.2 about different forms of associative thinking that so often give rise to creative ideas. This section considers a couple of them in more detail and explores further sources that may stimulate associative thinking.

Adaptation

If a problem is not in a completely novel area, a solution is often found by finding something similar and adapting it. Adaptation may be applied at the total product/system, sub-system, or component level. Most designers – including Moulton, Starley, Sanders and Ritchie – employ this common sense heuristic approach.

Checklist	Adaptation
1	Apply the technical principles of an existing product to a new design.
2	Copy the best elements or components of an existing product. Adapt or change elements that are unsatisfactory.
3	Create variants on existing designs e.g. different shapes, sizes, colours.
4	Use the design style or thinking of famous designers, e.g. Philippe Starck, to inspire a new design.

Exercise 14 Associative thinking

Supposing, like the designers discussed in Section 3.1.2, you wanted to design a portable bicycle. What questions might you pose yourself to find some new ideas for concept design?

Spend 10 minutes making a list of associative thinking questions in your workfile before looking at my response below.

My response

Below is my checklist of questions:

Checklist	Associative thinking
1	Do I know of a problem that is similar in any respect?
2	Is there an existing device used in a different application that could be adapted?
3	What other objects operate similarly (i.e. can be folded, collapsed or taken apart)?
4	Are there any devices that have a similar function?

Considering these questions provided me with the following products which function on principles that might be adapted or provide ideas for the portable bicycle design: umbrellas; baby buggies; folding doors; deckchairs; folding chairs; hand fans; telescopes; tripods, tents, textiles, sofa-beds, jigsaw puzzles.

Perhaps I could design a bicycle that opened like a pop-up tent or slotted together like pieces of a jigsaw?

Transfer, combination and analogy

Using lists of questions such as those above produces not just ideas for objects that might be adapted. It also helps the designer to think of possibilities for *transfer* of ideas between different fields of design and technology, for *combining* ideas and for thinking in terms of *analogies*.

Specific sources for associative thinking

As well as the general ways in which associative thinking can provide inspiration, there are some specific sources such as new technologies and materials, and socio-economic trends.

New technologies and materials

> I love the possibilities inherent in different materials … natural [organic] materials above all are incredibly inspiring.
>
> (Emmanuel Dietrich, French product designer, quoted in Fiell and Fiell, 2001, p. 131)

An important source of ideas is new technologies, materials and components. Trends in the technical or design evolution of products (e.g. the examples given in Block 2 Section 2.2) may also offer a stimulus. Designers should therefore try to keep up-to-date with technical trends and developments relevant to their field.

Exercise 15 Evolving technologies and materials

Returning to our energy-efficient vacuum cleaner example, what new technologies, materials or components, or trends in design, technology or manufacturing, might lead to an idea for a new concept?

Spend 10 minutes making a list of technical changes in your workfile before looking at my response below.

My response

Checklist Changing technology

1 Could the following be applied or provide ideas?

- new technologies (e.g. micro-electronics, 'intelligent' products);
- new materials (e.g. carbon fibre, organic or recycled materials);
- new components (e.g. flat 'pancake' shaped electric motors).

2 What trends in technology, design or manufacturing may be relevant?

- towards smaller, more powerful machines;
- towards more energy-efficient, less polluting devices;
- towards automatic or maintenance-free products;
- towards small production runs of customised products.

Socio-economic trends

> The creative process is one of observation of the world around us … we draw our inspiration as much from eating a sandwich in a local deli as from researching … materials.
>
> (El Ultimo Grito, quoted in Fiell and Fiell, 2001, p. 152)

In Block 2 and the DVD sequences on IDEO and Philips you saw that future thinking, scenarios and individual 'personas' are useful for product planning and design. Product designers are also often highly tuned to trends in society, lifestyles, design and fashion and this can provide a stimulus or inspiration for new concepts.

Exercise 16 Socio-economic trends

What social, lifestyle, design or fashion trends might be used to provide ideas to a designer of vacuum cleaners?

Spend 10 minutes making a list of socio-economic trends in your workfile before looking at my response below.

My response

A designer of vacuum cleaners might get ideas from the following trends:

Checklist	Changing society
1	*Society*: more working women; more single person households; changing male/female roles; more elderly and retired people.
2	*Lifestyles*: less time for domestic chores; many users of products in each household; homes with less space.
3	*Design* to create 'objects of desire' that appeal to the emotions; for improved usability; to reduce impacts on the environment.
4	*Fashion* e.g. in rounded shapes or pastel colours; in new types of floor covering.

For both learner and experienced designers, associative thinking is probably the most fruitful source of creative ideas and concepts. You should be on the lookout for objects and information from almost anywhere that might lead to a solution.

5.2.5 Discussion and teamwork

Another obvious thing to do when attempting to solve a problem is to involve other people. The involvement can range from an occasional discussion to tackling the problem in a team.

Most designing is a team effort. Teams can help to clarify a problem and generate some initial ideas through formal meetings or informal discussion. Many design teams use a loose form of brainstorming 'to toss ideas around' at the early stages of a project (I'll describe formal brainstorming later in Section 5.3 and on the DVD). When some initial ideas have been produced, meetings, discussion, and further brainstorming is a very good way of generating more ideas and developing the initial ones.

A useful way to involve other people in developing any piece of creative work is to *circulate some preliminary ideas or rough draft concepts to a few people with relevant experience*, and then to get their comments and/or meet to discuss and develop your initial ideas. Further versions are then circulated for comment and improvement until the work is finalised. This is essentially employing the powerful heuristic of 'successive approximation' or *iteration*.

By involving other people you can usually work more creatively than by keeping the problem to yourself. A danger is that you might expose your ideas to criticism at a stage when they are only partially formed and vulnerable. One way of guarding against this is to resolve to treat criticisms as opportunities for improving your ideas rather than as discouragement. Nevertheless, it is important to ensure that the people you discuss ideas with are not inclined to be over critical and are prepared to work on the ideas rather than wanting to knock them down.

> Discussion and teamwork, especially to facilitate iterative design, is valuable for both learner and experienced designers.

5.2.6 2D sketching and 3D modelling

Sketching and drawing

> Designers often engage in sketching not to record an idea that is not there yet but to help generate it. This is visual thinking.
>
> (Tovey *et al.*, 2003, p. 140)

Two-dimensional sketching is used by designers not just to represent and communicate the ideas and concepts they have envisaged in their heads, but to create, explore, evolve and evaluate those ideas in a form that is quick and easy to change.

Recall Philippe Starck's series of sketches when evolving his lemon squeezer (Figure 4 in Section 1) and Mark Sanders' sketchbooks when creating the Strida folding bicycle (Figures 51 and 52 in Section 3). These designers are employing *iteration* through 2D modelling.

That's why you should make rough sketches and diagrams when working on ideas and concept designs. Figure 66, for example, shows concept sketches by postgraduate car design students and Figure 67 is a sketch of a sweater produced by a knitwear firm's designer that might now be produced directly on a CAD system. It's often helpful to annotate your sketches, like the alarm clock concept sketches shown in Figure 18 in Section 1.

Figure 66 Concept sketches of car designs produced by postgraduate automotive design students

Figure 67 Rough sketch design for a women's sweater

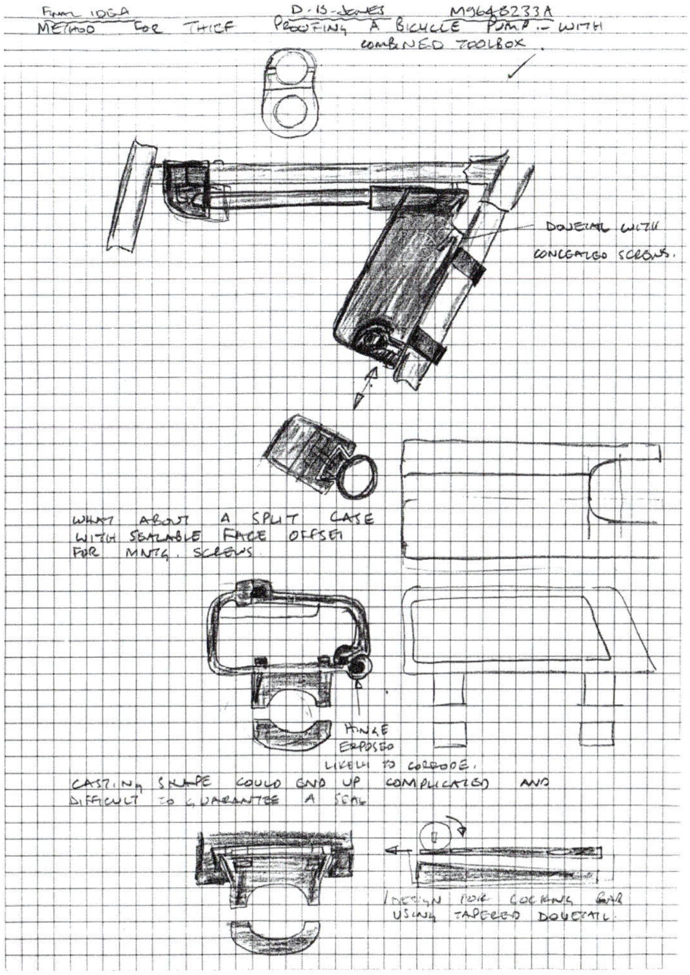

Figure 68 Annotated sketches by an OU student exploring ideas for a combined toolbox–bicycle pump–security device

Remember you're using sketches and drawings to aid *your* thinking and develop *your* ideas, so the standard need not be high. Recall the student design team's simple concept sketches of the Splashguard (Figure 60 in Section 4), while Figure 68 shows sketches by a student of an earlier second level OU design course when working on improvements to the bicycle.

3D modelling

Three-dimensional modelling is also vital in order to create, evolve and evaluate concepts and designs. 3D modelling includes principle-proving models to test feasibility, such as Sanders' wire models to test and refine two folding bicycle configurations (Figure 69), and scale models, for example Moulton's models of his small wheel bicycle (Figure 39 in Section 3). Card and foam are materials often used to make simple mock-ups and models, as described in the *Modelling Workbook* (e.g. Figure 27 showing foam models of a video camera and Figure 29 of the cardboard model of a seat.)

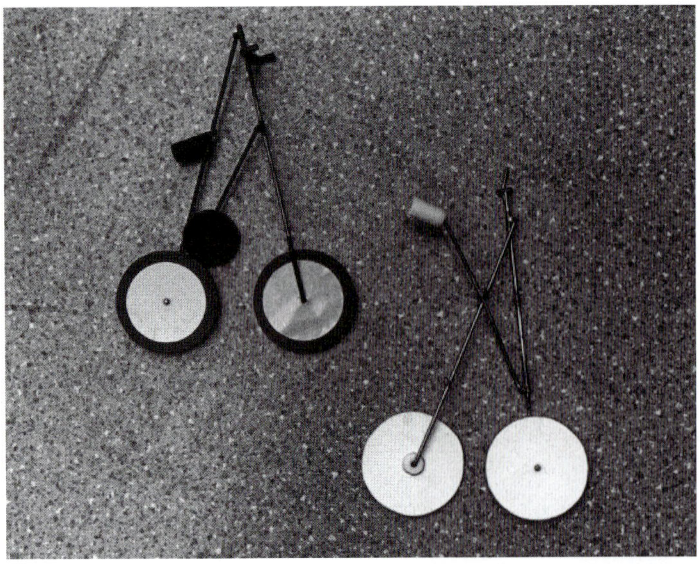

Figure 69 Wire and card models used by Mark Sanders to present, test and refine two possible folding bicycle configurations, the X-shape and triangular forms

2D and 3D modelling are basic creative skills for both learner and experienced designers. You can learn and practise some of these skills using the *Modelling Workbook*. If you haven't already done so, you should do the exercises in Sections 4 and 5 of the *Modelling Workbook*. This would also be a good point at which to do some preparation for the sketching involved in the Block 3 TMA (see the *Assignment Booklet* and *Block 3 Supplement* for details). Before starting serious work on the TMA you should also look at Activities 13–15 in Section 6, and Figures 97 and 98 in Section 7 of the *Modelling Workbook*.

You can see once again how important 2D and 3D modelling are in the designing process in two DVD video sequences that show how two design students each carried out a creative design project for the Dyson Design Prize 2001.

Before starting Section 5.3, or at the end of Section 5, look at the DVD notes 'New designers' in the *Block 3 Supplement* and view the DVD sequences 'Pushchair', about the conception and development of a new design of baby buggy (Figure 70(a)), and 'Battery charger', about the development of a human powered battery charger (Figure 70(b)).

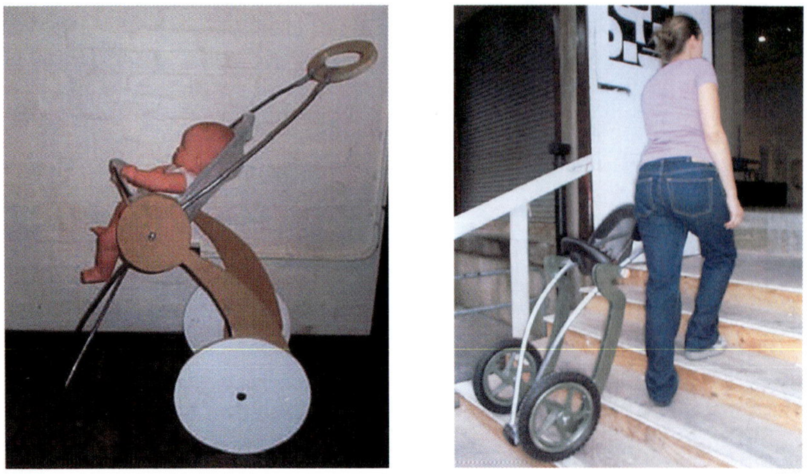

(a) The 'Baby Bug' (Sarah Liddell)

The second test rig (left) and the final prototype being pulled up some stairs (right)

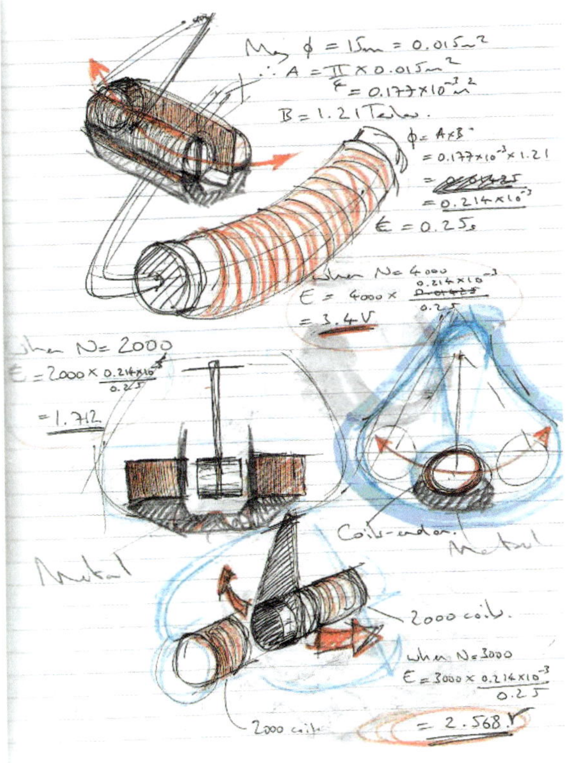

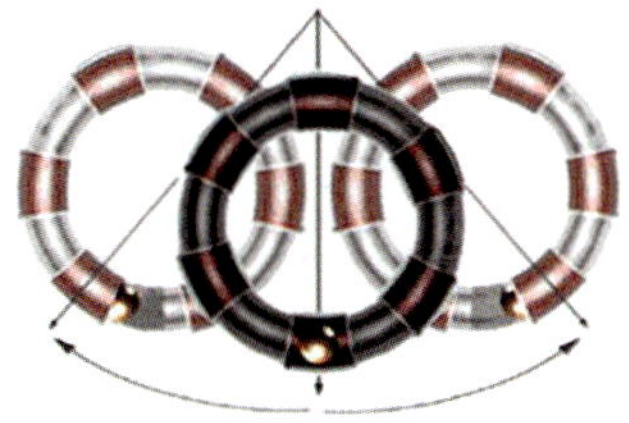

(b) The 'Electric doughnut' human-powered battery charger (Justin Floyd; first prize)

Sketches of the second concept (left) – a pendulum magnetic induction system that led to the final doughnut-shaped design (right).

Figure 70 Dyson Design Prize 2001, two post-graduate industrial design students' entries

5.3 Creative problem solving techniques

Experienced designers' approaches to creativity and concept design can be very effective, as should be evident from the examples in the block so far. As Section 5.2 above outlined, *learners should also use the approaches of experienced designers*. In most cases applying these approaches – immersion, association, discussion, sketching, etc. – should produce a satisfactory solution.

But what if you are a learner designer and don't know how to proceed, haven't time for immersion and need some ideas quickly? *Creative problem-solving (CPS) techniques*, such as brainstorming or morphological analysis, have been developed to help in such situations. Several CPS techniques are similar to systematic design methods, or can be applied within a systematic design approach. (There's a specific technique also called CPS, here I'm using the term generically.)

To understand how CPS techniques fit into a systematic design approach, look back at the model of the design process introduced in Section 1 of this Block (Figure 15).

This describes the design process as repeated activities of problem *exploration* (or design *specification*), idea/concept *generation*; and concept *evaluation/selection*. There are CPS techniques to help with each of these activities.

CPS techniques have mainly been used by managers, advertisers and marketers and – apart from visual brainstorming – are only rarely used by experienced designers. This is because they seldom produce any worthwhile, practical ideas or designs that wouldn't have come from heuristic approaches. *However, CPS techniques can be useful to learner designers.* This is because they formalise certain heuristic methods used intuitively by experienced designers.

Numerous CPS techniques have been devised, but the majority are intended primarily for *idea generation*. These enable learners to explore a particular solution space more widely than they might otherwise have done. Experienced designers have already done much of this exploration during their working life. My own experience is that idea generation techniques are most useful when you have defined a specific question to which you want alternative answers. For instance, if you were working on the design of a powered vehicle, the techniques are a useful way of quickly generating alternative ideas for power sources. So, in design, CPS techniques are generally of more practical value to generate ideas at the *sub-system* and *component* levels than at the total product/system level.

Although CPS techniques are often presented as rules to follow, once learned you shouldn't apply them rigidly. As Nigel Cross (2000) says they are best viewed as *'life-jackets' to keep you afloat rather than as 'strait-jackets' to be forced into*. Often the main benefit of using CPS techniques is they help you to 'flex' away from your usual style of thinking. For example, if you tend to be a convergent thinker they can help you be more divergent, or vice versa.

> In this section I shall outline a few problem exploration, idea generation and concept selection techniques. You are *not* expected to learn the techniques from the outlines below: just read them to get a feel for what each technique can do. Instructions of how to apply the techniques are provided on the DVD. There is no need to look at the DVD material until the end of this section.

5.3.1 Problem exploration techniques

Some problem exploration techniques are similar to the essential function/performance specification approach, which is explained in Section 5.3.2, so I won't cover them here. Instead I've chosen mind-mapping, a useful general-purpose technique for problem exploration and idea classification.

Mind mapping

A mind-map (or spray diagram) is a simple and useful way of identifying the different elements of a problem and the relationships between these elements, after you have immersed yourself in it. Mind mapping often also produces initial ideas for solutions. It is something you may have learned from other OU courses or used in other situations.

Mind mapping is also valuable for *classifying and selecting ideas* such as those produced by idea generation techniques. Its use in classifying and selecting ideas in illustrated in Section 5.3.3.

5.3.2 Idea generation techniques

How do the idea generation techniques work?

All of the idea generation techniques work on a limited number of principles:

First, they stimulate an individual or group to conduct a *divergent search* for ideas and to consider odd or zany ideas that otherwise might have been excluded. Second, the techniques often require the individual or group to '*suspend judgement*' on the quality or viability of the ideas so as not to converge prematurely and kill off half-formed ideas. (They thus differ from experienced designers' approaches by diverging widely *before* converging, rather than switching frequently between divergence and convergence.)

Some techniques work by providing words, images, etc. that stimulate an individual or group to come up with ideas spontaneously. These are usually called *intuitive techniques* (with some similarity to the heuristic methods of designing discussed above).

Other techniques work by breaking the problem into parts, identifying possible sub-solutions and then generating overall solutions by permutation and combination. These are usually called *logical* or *systematic techniques* (similar to some systematic methods of designing discussed above).

These two basic types of idea generation themselves operate in two ways:

1 *free association* in which new ideas are stimulated in the mind arising from the experience and environment of the individual and by ideas produced by other people;

2 *forced relationships* where new ideas are produced in the mind by making people consider connections between related or unrelated ideas.

Combining these categories gives four types of idea generation technique, as shown in Table 1. Within the matrix are the techniques you will find outlined below, and (in brackets) some approaches you've met earlier that fit into this classification.

Table 1 Classification of idea generation techniques

	Intuitive	Logical/systematic
Free association	Brainstorming/brainwriting (Checklists)	(Essential function/ performance specification)
Forced relationships	Random stimuli (Associative thinking)	Morphological analysis

(Based on Geschska *et al.*, 1982; Van Gundy, 1988)

Intuitive techniques

Brainstorming/brainwriting

Brainstorming is the best known of the CPS techniques. It's simply a method of getting a small group of people (ideally four to seven) to generate a lot of ideas in response to a defined problem or question.

'Brainstorming' is often used to mean a freewheeling discussion, but the brainstorming technique has definite rules – as shown on the IDEO and 'Brainstorming/brainwriting' DVD sequences. The most important rule is *'encourage wild ideas'* of which 'no criticism is allowed' even of the craziest or apparently silly idea. Evaluation of the ideas comes later. Another important rule is that participants should use the ideas of others to spark off their own ideas and *'build on the ideas of others'*. In design visual brainstorming with rough sketches of ideas is generally more effective in generating and communicating the idea than using words. The outcome of a brainstorming session is therefore perhaps 50 or more ideas in sketches and/or words for tackling a given problem which can then be classified and evaluated.

For concept design a more structured form of brainstorming generally produces more worthwhile ideas. Here brainstorming is used to generate concept design ideas, but with the constraints of the brief or specification always in mind – the so-called 'criteria-cued' mode (Pugh, 1991). Alternatively, brainstorming is used to generate ideas for similar products, technologies or situations (analogies), which may then be used to provide stimuli for new concept designs.

Brainwriting is similar to brainstorming, except that ideas are written down instead of spoken aloud. It is therefore suited to individual as well as group idea generation. There are several variants of group brainwriting, such as 'Method 635' in which six people each write down three ideas in response to a given problem in five minute intervals. Each person's ideas are then passed to another participant and the process is repeated using the first set of ideas to spark off more ideas or to refine the initial ideas.

Often the best results are obtained by beginning with individual brainwriting and then sharing the individual ideas to start group brainstorming (or to elaborate the individual ideas through informal discussion).

Random stimuli

Random stimuli work rather like analogies to stimulate the mind to think in new ways. A random stimulus may be introduced in a variety of ways. A simple one is to pick a random word from a book, or picture from a magazine, and to think about the word or picture in conjunction with the problem to be tackled to see what ideas it suggests.

Suppose the problem is to devise a way of fastening cupboards to a wall so they are properly aligned horizontally and vertically. The word 'bird' is chosen at random. Thinking about the problem in relation to the characteristics of a bird: flying, feathers, claws, etc. suggests several ideas. For example, birds flying suggested to one individual the idea of an air bag placed underneath the cupboard which by pumping air in or out brings the cupboard into the desired position where it can be easily screwed to the wall (Geschka *et al.*, 1982). Similar devices are already used for positioning very heavy objects.

In theory the more remote the stimulus from the problem in hand the more novel the ideas that might be stimulated. In practice it is often objects or images that are already connected consciously or sub-consciously to the problem that work best. Hence other random stimuli techniques involve going for a walk and using something in the surroundings that catches your attention as the source of ideas.

The use of random stimuli may seem silly or far-fetched; but they can and do often work, although of course the ideas they generate are not guaranteed to provide a workable solution to a problem.

Thoughtpath Problem Solver™

This is a computer program that uses random stimuli methods to help you explore new ways of solving a problem and to 'think outside the box'. *Thoughtpath* is included on the DVD and there are further details in the *Block 3 Supplement*, but there is no need to attempt to use this software yet.

Logical/systematic techniques

These techniques depend on breaking a problem into parts in order to generate solution ideas, and therefore tend to appeal to engineers and others used to a more analytical approach.

Morphological analysis

Morphological analysis is perhaps the most widely used of the logical techniques. It involves analysing a class of products into its main features or functions (called 'parameters') and then considering different combinations of possible ways of meeting the parameters. Morphological analysis can generate a very large number of solution concepts.

My suggested approach in answering the alarm clock exercise in Section 1.4 was a simple example of morphological analysis. In that case the parameters were the 'power source', the 'time indicator' and the 'alarm system'. Different ways of meeting the parameters were identified and combined to generate concepts for different types of clock. The Splashguard team also used a form of morphological analysis (Box 3 in Section 4.1).

Now consider another example, the different concepts for wave-power devices in Figure 13 from Section 1. Morphological analysis can be used to identify other possibilities. The essential features or functions of any wave-power device is that there has to be a *working surface* on which the waves act, a *balancing force* to hold the device in place, and a way of concentrating the *wave energy* and transferring it to a turbine or other system for converting it into, say, electricity. Thinking of alternative sub-solutions, or ways of meeting each of these parameters produces the morphological chart shown in Table 2.

Table 2 Morphological analysis of wave-power devices

Parameter (function)	Ways of performing function		
Working surface	Flexible sheet	**Rigid solid**	Air surface
Balancing force	Fix to sea bed	Inertia	**Moving structure**
Energy transfer	Air flow	Sea water	**Hydraulics**

(Adapted from French, 1988, p. 295)

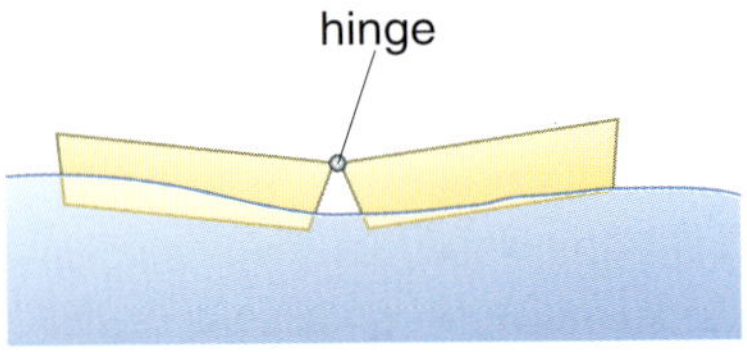

Figure 71 Hinged raft wave-power device

The two solid halves of the raft follow the wave contours and the wave energy is transferred to turbines by hydraulic pumps operated by the relative motion about the hinge

The different combinations of sub-solutions in the chart generate 27 ($3 \times 3 \times 3$) possible types of wave-power device, including all those shown in Figure 13. Another concept suggested by the combination in bold on the chart is the hinged raft type (Figure 71). The working surface comprises the two solid halves of the raft which follows the wave contours, the balancing force is provided by a structure of linked rafts, and the wave energy is transferred to turbines by hydraulic pumps operated by the relative motion about the hinge.

5.3.3 Concept selection techniques

Following idea generation it is, of course, necessary to select one or more promising ideas or concepts for further development. This can be done using judgement e.g. using the *go/no-go* method outlined for the Splashguard in Section 4.1.5, or using one of the CPS techniques for evaluation and selection. Most depend on *classifying* ideas to find patterns and/or using checklists of *criteria* to choose between alternatives. (The criteria are usually those given in the design brief or specification, as discussed in Block 2.)

Classifying and selecting ideas

This involves grouping ideas produced by one of the idea generation techniques in order to find clusters of similar ideas, perhaps at the same time filtering out impractical ideas and generating some additional ideas. The ideas may be recorded on cards, Post-It notes or flipcharts and the classification can be done individually and/or in a group. Alternatively the ideas may be clustered on a *mind-map* (see Section 5.3.1 above and Figure 72).

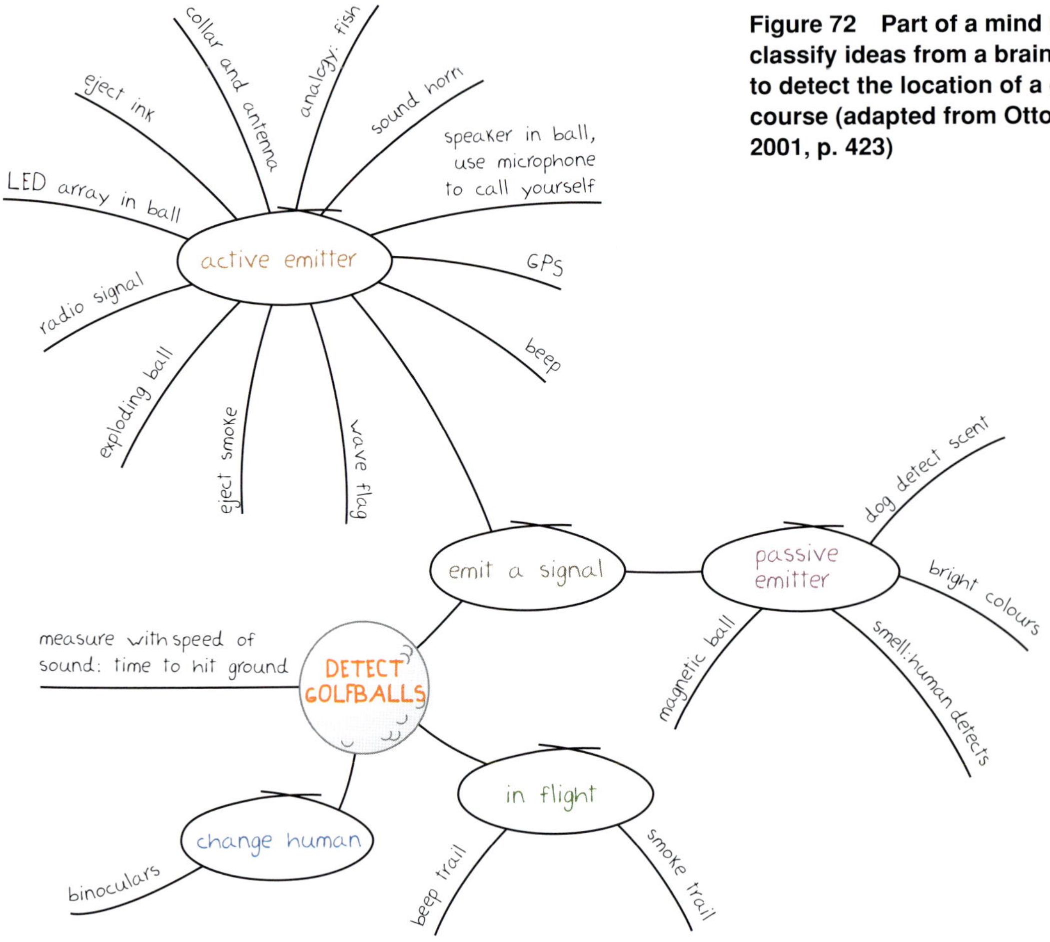

Figure 72 Part of a mind map to classify ideas from a brainstorm on how to detect the location of a golf-ball on a course (adapted from Otto and Wood, 2001, p. 423)

In a group, the idea clusters may be narrowed down to a few favourites by voting, for example, each member can award each cluster up to three stars and the idea(s) with most stars survive. Assuming the evaluators have the expertise to make reasonable choices, the method produces a shortlist of worthwhile ideas for investigation and development.

Selection matrix

A selection (or evaluation) matrix was used to select the best technical concept for the car horn case study in Section 1 (Figure 17). The technique involves drawing up a matrix to systematically score each concept relative to an existing (or favourite) design (the *datum*) using criteria from the design specification. The criteria may be weighted according to their relative importance. The concept(s) with the highest relative score are selected for development.

Of course the selection process does not end there. Often it will be necessary to go back to the list of ideas and concepts, or even generate new ones, if on further examination a selected concept turns out to be technically infeasible, not economically viable, or otherwise impractical.

Earlier ideas may offer solutions to the new problems that are inevitably thrown up as a concept is developed. So no idea should be completely forgotten.

Your workfile is particularly useful for recording how your ideas develop when answering the Block 3 TMA.

The DVD material for Block 3 includes instructions and interactive exercises to help you learn the CPS techniques outlined in Section 5.3. This is a good point at which to use the DVD to try the CPS techniques. Full details are given in the *Block 3 Supplement*.

Key points of Section 5

Experienced designers and teams often use a heuristic approach to create high quality concept designs.

Learner designers should in the first instance attempt to *use the approaches of experienced designers*. But they may also need the framework and aid provided by *systematic design methods* and *creative problem solving (CPS) techniques*.

Key heuristic approaches for both experienced and learner designers are:

- preparation by *immersion* (in the problem and in existing designs) – gathering information, reviewing existing solutions, producing initial concepts;
- identifying a *problem frame*;
- *associative thinking* – adaptation, transfer, analogy, combination;
- *discussion and teamwork*, e.g. to aid *iterative* improvement of draft concept designs;
- *2D sketching* and *3D modelling*, also to aid *iterative* improvement of draft concept designs.

Useful and relatively easy to learn *design methods* and *CPS techniques* for different elements of designing include:

- *problem exploration/specification*: essential function/performance specification; mind-mapping;
- *idea generation*: brainwriting/brainstorming, random stimuli, morphological analysis;
- *concept evaluation/selection*: classification/selection, mind-mapping, selection matrix.

(Learning outcomes 1.2–1.4, 2.2–2.4, 3.1–3.4, 4.1 and 4.2)

6 Creativity is not enough

So far you've learned about the importance of creativity throughout the designing process, focusing on the concept design phase. You'll have realised by now that being creative is not enough to ensure that a new product meets the needs of its users, succeeds in the market, or satisfies other requirements such as reducing environmental impacts. Indeed it can be argued that designers, encouraged by design schools, sometimes try to be *too creative*, producing innovative designs when there are perfectly acceptable existing products that could be improved.

As noted in Section 1, in this block I have focused more on innovative design than the less risky and more common activity of 'variant design' aimed at improving existing products. In this section therefore I'll briefly review some of the factors that determine whether an innovative product design becomes a successful *innovation* – namely, that it is first *launched* onto the market and then *diffuses* into widespread use.

On the DVD James Dyson talks about the problems he experienced in getting his Cyclone cleaner into production and onto the market. And on the sequence about the Strida, Mark Sanders discusses the market and manufacturing issues he faced that led him to design the improved Strida 2 and 3 folding bicycles. Further information on these sequences is given in the *Block 3 Supplement*. If you haven't already viewed these sequences as part of your study of Section 2.3 (Dyson) or Section 3.1.2 (Sanders), you should do so during your study of this section.

6.1 Obstacles to innovation

The obstacles to innovation are those which may prevent or delay an invention or a new design reaching the market. There are many such obstacles, the most important of which are *technical*, *financial* and *organisational*.

6.1.1 Technical obstacles

There are a number of different types of technical obstacles. These are outlined below:

1 A clear technical obstacle to innovation is if an idea is scientifically impossible. Ideas or concepts, however creative, which violate scientific laws are clearly not going to get very far. This does not prevent the Patent Office from regularly receiving designs for perpetual motion machines and other impossible inventions.

2 Some designs or inventions are infeasible because of the existing state of technology, materials, components or knowledge. They may eventually reach the market, but will be delayed. For example, Thompson's original pneumatic tyre was technologically in advance of its time and innovation was delayed until Dunlop developed his tyre. Moulton's small-wheel bicycle would have been delayed, or may never have reached the market, had the Dunlop company not been able to provide him with high-pressure cycle tyres.

3. Another common technical obstacle is when a new concept fails to reach the market because of problems in the development or detailed execution of the design, e.g. to make it sufficiently reliable, attractive looking or economical to manufacture. Sometimes the technical difficulties of producing something in sufficient volume to be financially viable may prevent innovation.
4. An obstacle related to the above one arises where the technical capabilities of the individual or organisation that is attempting to develop a design, or its production process, is not up to the task.

While the latter two are obstacles to be overcome, rather than absolute barriers to innovation, they probably account for many more failures than the first two. The failure of the many weird and impractical designs produced by nineteenth century cycle designers was partly due to the fact that most were blacksmiths or mechanics rather than trained engineers. Section 3.1 stressed the necessity for at least some technical expertise to successfully design even relatively simple products such as bicycles.

6.1.2 Financial obstacles

Generating ideas, inventions and concept designs is usually the least expensive part of the innovation process. Developing those ideas into a practical device, investing in tooling and providing manufacturing facilities is much more costly. Obtaining finance for the development of an idea and for production and marketing is therefore a major obstacle to innovation, particularly for the independent inventor-designer.

Recall the example of the Brompton folding bicycle (Section 3.1.2). Its initial production could only be in small batches, each funded by the sales of the previous batch, and mass manufacture was delayed by five years because of lack of finance. This was compounded when an outside manufacturer stopped making an essential component that was too costly to make in-house.

Innovative products, like the Brompton, need to sell in sufficient volume to recover the money invested in their development. If future sales are unlikely to repay the loans or investments for development, manufacture and marketing within a given time, the new product may never be financed. Or, as in the case of many novel designs made by small enterprises, the innovation may go into small-scale production for short periods before cash flow problems halt the business.

6.1.3 Organisational obstacles

Innovation involves disturbance and change and therefore tends to be resisted by organisations or the individuals within them. This accounts for the unwillingness of many established businesses to accept radical new ideas and designs, especially those from outside the organisation itself. Dyson, when he first approached major vacuum cleaner makers with the Cyclone, and Moulton, when he first tried to get Raleigh to make his small wheel bicycle, strongly encountered the so-called NIH ('not invented here') syndrome.

Moulton recalled:

> ... one of the [company's] officials scrubbing [dragging] the bike *sideways* along with an expression of hatred. He must have regarded it as a threat to the stability he found in the familiar design of the classic bicycle.
>
> (Moulton, 1979, pp. 41–2)

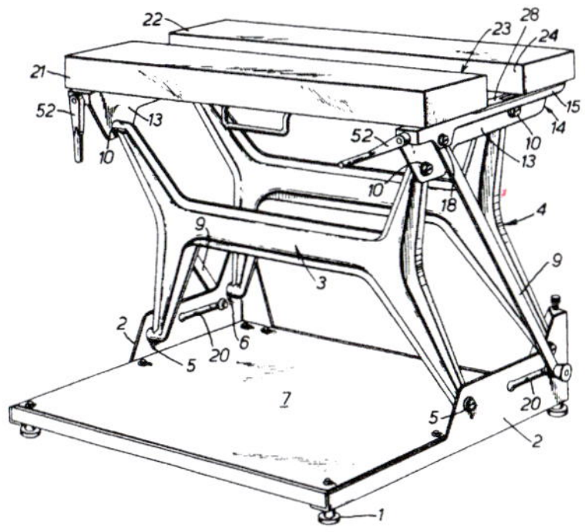

Figure 73 Patent drawing of the Mark I 'Workmate' invented and designed by Ron Hickman

Hickman had to make and sell this version himself before existing manufacturers were convinced that his innovation was commercially viable.

A well-known example of established organisations failing to see the potential of an innovative product is the Workmate home workbench (Figure 73). None of the manufacturers of tools and do-it-yourself equipment that its designer, Ron Hickman, approached were willing to take on the Workmate and, like Dyson and Moulton, Hickman had to make and market his innovation himself. Again, as in the cases of the Cyclone and the Moulton bicycle, only after the innovative product proved to have a substantial market did manufacturers become interested. Black and Decker agreed to licence the Workmate from Hickman - it was to become one of their most successful products, selling over 10 million units in the first ten years.

A lesson here is that existing manufacturers are usually very reluctant to take the *risks* involved in innovation. But if the innovation proves successful, established manufacturers may either try to licence it or to overcome any patents and copy it. There are now many variants of the Workmate. And 'bag-less' vacuum cleaners from long-established manufacturers, using either conventional vacuum or cyclonic technology, now compete with the Cyclone. (This is despite Dyson winning court cases against Hoover in 2000 and 2001 for infringing his patent.) But sometimes established manufacturers resist a new 'disruptive' innovation for so long that whole new industries based on that innovation have emerged and the established firms may decline or go out of business (Christensen, 2000).

6.2 Obstacles to diffusion

So far I have considered the obstacles that influence whether a new product becomes an *innovation*: i.e. reaches the stage of being launched on the market for the first time. But once an innovation has been introduced, there are many factors that influence the rate at which it sells or spreads into use. These are the obstacles to *diffusion*, which depend on several factors: the nature of the innovation itself, the market for the innovation, and of the individual or organisation promoting the innovation. I'm now going to examine why some innovations are more successful than others.

6.2.1 Characteristics of the innovation

In his classic book *Diffusion of innovations* Everett Rogers (1983) gives five characteristics of an innovation that influence how rapidly it is likely to diffuse. I'll consider four of these in turn (the fifth characteristic 'trialability', meaning whether an innovation may be tried out before purchase, is less relevant to my discussion).

Relative advantage

Unless an innovation is perceived to offer *advantages relative to comparable products* it is unlikely to succeed. The more highly priced the innovation is relative to comparable goods or services the greater the advantages have to be.

Designers and inventors often ignore this obvious fact. They are so caught up with their new product that they fail to see how it will be perceived by potential customers when they compare it with what else is available.

The important word here is 'perceived' advantage. As Block 2 showed, different people have different mixes of demands and wishes they want a product to satisfy – performance, reliability, features, usability, attractive appearance, safety, low environmental impact, etc. To diffuse, an innovation will have to offer relative advantage(s) to at least a segment of the market. The advantage may be as much a matter of fashion and image as improved performance, features or price. The fashionable image of mountain bikes created by their bright colours, rugged appearance and the idea that a mountain bike (like a four-wheel drive car) offers the freedom to go almost anywhere is one reason for their great popularity. Another reason is that their strong frames and broad tyres are perceived by buyers as offering advantages over conventional bicycles for negotiating roads with kerbs, potholes and broken glass.

It isn't enough to have a relative advantage when an innovation is first launched. Indeed when first introduced, innovations often function less well or are more expensive than established products, but offer something that some customers value. Digital cameras, for example, initially produced poorer images than conventional photography, but enabled users to electronically manipulate the images. The first mountain bikes were heavy-duty 'clunkers' assembled from available components that enabled a few Californian enthusiasts to ride down mountains.

To maintain an advantage it is usually necessary to *continuously improve* the product. Depending on the market aimed at, this improvement may be in performance, features, image, price, etc. Digital cameras now equal conventional photography for image quality, have come down in price, and are smaller, lighter and more fashionable (remember the example on specifying the 'benefits' of a new digital camera in Block 2). With Kodak's announcement, in January 2004, that it would cease to market film cameras in the UK and US it looks like digital cameras will become the dominant technology in the market, with film cameras perhaps relegated to niche/specialist markets.

Figure 74 Mountain bike with front and rear suspension

The frame has been reconfigured to accommodate the suspension.

After initial development by their enthusiast users, mountain bikes are now sophisticated products, most with aluminium alloy frames and suspension systems on front and/or rear wheels (Figure 74). Mass production in the Far East using advanced manufacturing technology has made mountain bikes affordable. As you've seen, Dyson has continuously improved and expanded the range of his Cyclone cleaners, offering new colours and designs, improved features and cyclonic technology. Moulton likewise has continuously improved and developed his bicycle designs.

Compatibility

To succeed an innovation has to be perceived as being compatible with the existing values, skills and past experiences of potential customers and users. Many innovations fail because they are inconsistent with such values, skills and experiences. They may also fail due to incompatibility with existing products and systems.

Section 3.1.2 noted that, although there are numerous folding bicycles on the market, few have sold in any volume. The idea of having a bicycle you can carry in a car, on the train, etc. is appealing in theory. But in practice few people in industrialised countries are willing to abandon the convenience and comfort of their cars, or bother with the effort and skill of cycling, even if it helps reduce pollution. Cycling may also be incompatible with the need to dress smartly, carry luggage and park securely. However, perceptions of compatibility can change with changing circumstances. For instance, increasing costs associated with using cars (such as road tolls or congestion charges) and the increasing costs of car parking at railway stations etc. may lead to larger markets for folding bicycles in the near future.

Complexity

The complexity of an innovation is the extent to which potential customers perceive it as being difficult to understand and use. In general, the greater the perceived complexity of an innovation the less likely it is to succeed on the market.

Perceived complexity is another probable reason for the limited popularity of folding bicycles, as they are likely to be viewed by many as being difficult to fold and unfold, and to ride, carry and store. Sanders attempted to address this obstacle by designing the Strida to look and be as simple as possible to fold, unfold and use.

Observability

Observability is the extent to which adopting an innovation is visible to others. In general the more 'visible' an innovation is, and the more it offers its purchaser or user the opportunity to display their new acquisition, the more likely it is to succeed. Thus, the Apple iPod digital music player (Figure 30 in Section 2) became a very successful product not only because of its performance and usability but for its highly recognisable and desirable design. Likewise, mobile telephones have been an extremely successful innovation because, as well as providing the unique advantage of being able to contact people and receive messages while on the move, they offer the owner numerous opportunities to display ownership of a product featuring the latest design and technology. In contrast, a 'non-observable' innovation, such as a fuel-efficient car engine, would tend to diffuse less rapidly. In such cases, manufacturers sometimes try to indicate the innovation's presence with logos and badges on the product.

But although high observability is normally associated with rapid rates of diffusion, as you will see shortly, in certain cases potential users may feel they would be embarrassed by displaying their purchase and the opposite will apply.

6.2.2 Characteristics of the market

Apart from the characteristics of the innovation itself, the nature of the market into which it is to be sold has a major influence on diffusion. Some markets contain a high proportion of individuals who will buy novel products (so-called *innovators* and *early adopters*). But, typically, the majority of potential buyers and users of an innovation are conservative and reluctant to adopt anything new. 'Consumer resistance' can be extremely frustrating for the inventor, designer or innovating firm. Thus, at first Dunlop had great difficulty in persuading cyclists to buy pneumatic tyres – indeed the innovation was initially ridiculed by the cycling world. Pneumatic tyres only became accepted after their superiority over solid and cushion tyres had been clearly demonstrated in cycle races.

6.2.3 Characteristics of the innovator

The nature of the innovator, whether an individual or an organisation, can have a major influence on whether an innovation becomes successful. James Dyson is a successful innovator because, at the same time as thinking of ideas for new designs, he considers their potential for manufacturing and marketing as commercial products. He is also a great publicist for his products.

Likewise, innovating firms are only able to succeed in selling a new product if they are able to market it effectively and overcome any resistance from competitors. You saw how Moulton was forced to sell his bicycle firm to TI-Raleigh when it and other manufacturers introduced lower priced small-wheel bicycles without suspension. Sometimes even major companies can fail with an excellent innovation if they are up against competitors with superior market power or marketing ability. In personal computers, even with what many view as a superior product, Apple has had to struggle against the market power of firms offering IBM-compatible PCs equipped with Microsoft software.

6.2.4 Regulation, legislation and policy

So far I have outlined some of the many factors that can prevent a new product being developed into an innovation and diffusing into widespread use. There is one major factor that I have left out, namely the influence of regulation, legislation and government policy. These can affect both the creation of innovations and the rate at which they diffuse.

For example, as you will see in a moment, the introduction in the UK of new regulations concerning electrically-assisted cycles stimulated several manufacturers to develop such machines. Similarly, environmental standards, regulations and laws are stimulating car manufacturers to develop alternative fuel vehicles, such as hybrid petrol-electric cars (Figure 75(a)) and fuel cell powered vehicles, and even concepts for radical new transport systems that combine electric bicycles and electric cars (Figure 75(b)).

(a) Honda Civic petrol-electric hybrid vehicle

Figure 75

(b) Honda Intelligent Community Vehicle System

This comprises four concept design prototypes – an electrically assisted bicycle, two short distance personal electric vehicles and a small electric car. The system is designed for shared community use within an urban area
(Fuad-Luke, 2002, p. 136)

6.3 Three examples of innovative products

Let's look now at how what you've just learned may be used to explain, or to predict, the market success of some innovative products.

6.3.1 The Freeplay clockwork radio

In Block 2 Section 4 the Freeplay clockwork powered radio was introduced as an example of sustainable product design. The original Freeplay radio launched in 1995 has been very successful, selling in volume around the world and making its inventor, Trevor Baylis, into a minor celebrity and millionaire (Figure 76). Although it's hailed as an ecological product because it eliminates the need for mains electricity or batteries, it was not originally created for environmental reasons. I'll consider the environmental credentials of wind-up radios at the end of this case study.

Figure 76 Trevor Baylis in his workshop with an updated Freeplay clockwork radio

Among his many interests, Baylis's passion has been inventing, especially products to help the physically disabled, and in the 1980s he created and developed a range of products for the disabled. Then in 1991 he was watching a TV documentary about AIDS in Africa and learned that in some regions radio was the only means of mass education that could help reduce the spread of AIDS. But the need for batteries or electricity made radio access expensive or difficult. That night he got the idea for a wind-up radio – he has said from dreaming about sitting

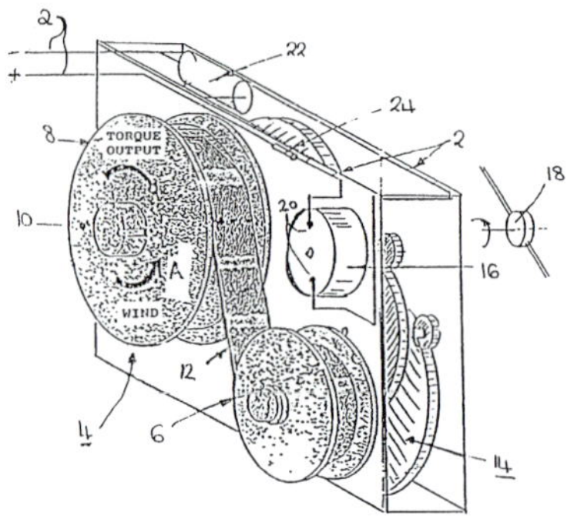

Figure 77 Patent drawing of the clockwork radio filed November 1991

The constant force spring is the band of steel wound onto spools.

in the African bush listening to a wind-up gramophone (*associative thinking*). The next day he went straight to his home workshop to make a mock-up to test his idea – an electric motor turned in reverse by a hand brace to act as a dynamo, connected to an old transistor radio (a *'principle-proving model'*). Baylis experimented with a number of concepts (*divergent thinking*) before designing a prototype clockwork radio powered by a constant force spring, gearbox and generator that played for 14 minutes with a two-minute wind (Figure 77). He applied for a patent and approached potential manufacturers. There then followed a four-year period of design and development.

Overcoming the obstacles to innovation

During this period Baylis experienced the many obstacles to innovation discussed earlier. His prototype would only play for a limited time (a *technical* obstacle). None of the manufacturers he approached were interested (*organisational* obstacles). And of course he needed money for development and production (*financial* obstacles).

He was fortunate to overcome these obstacles as a result of demonstrating the prototype on the BBC TV programme 'Tomorrow's World'. A South African accountant, Chris Staines, saw the programme, recognised the potential of a clockwork radio, and visited Baylis with a business proposal. Staines and a business partner then went to Africa to research the market and get reactions to the prototype. The battery-free prototype interested potential users, but to satisfy the African market it would have to be bigger and play longer and louder (*market research*, as discussed in Block 2).

The partners managed to get funding for product development from a variety of sources, including a South African foundation, which had the aim of establishing a manufacturing plant in Capetown that would employ disabled people. A consortium, including specialist spring and plastics manufacturers, a university engineering department, and a firm of design consultants, turned the prototype into a marketable product (QED, 1995).

Design and styling of the Freeplay was a particular challenge (Figure 78). It had to house the mechanical and electronic components being redesigned to provide the loud sound wanted by users; be robust enough to withstand hard use (even being dropped by parachute by aid agencies); appeal to African users – who generally prefer large, ostentatious-looking radios – while not deterring Western consumers. It had also to be designed for manufacture in a factory that would employ disabled workers. The final design of the original Freeplay is shown in Figure 79. Since the original clockwork radio, a range of new and improved Freeplay wind-up radios for European, US and other markets have been made, but in 2001 production was transferred from South Africa to China. A new design, the Freeplay 'Lifeline', has been developed, by the company's non-profit subsidiary the Freeplay Foundation, for educational and humanitarian use in Africa and other developing countries.

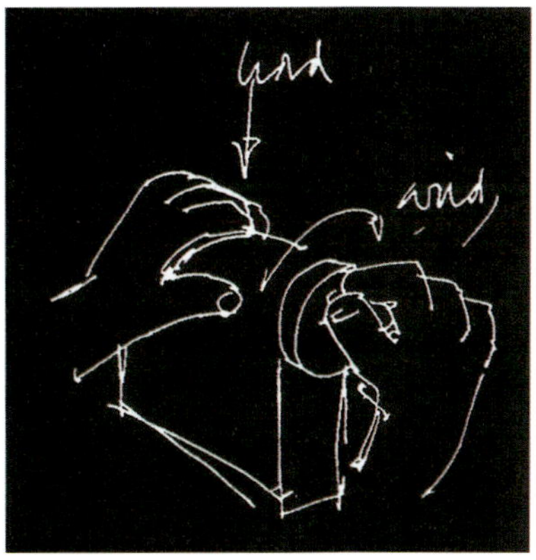

Figure 78 Styling sketches and prototypes of the original Freeplay radio (Evamy, 1995 p. 52)

(a) Final design of the original Freeplay radio

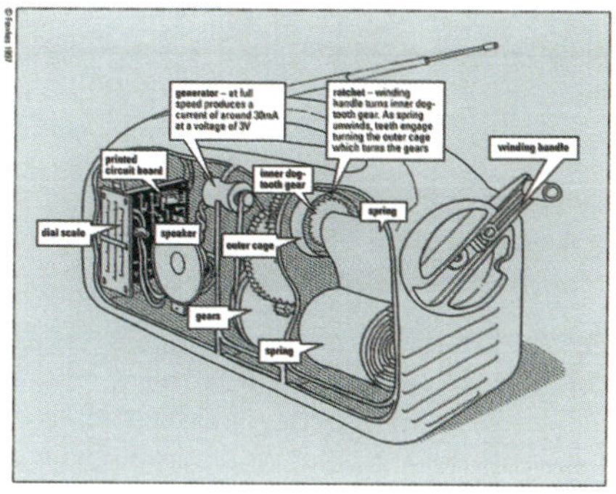

(b) Cutaway drawing of the original Freeplay radio (Chick, 1997, p. 55)

Figure 79 The original Freeplay radio

Overcoming the obstacles to diffusion

Why has the Freeplay been so successful? Let's look first at how it satisfies Rogers' diffusion factors:

Relative advantage. The Freeplay offers a unique advantage relative to other radios: it requires no batteries or mains power. This is likely to be appeal to some Western consumers attracted to the idea of a human-powered, 'green' product that will work anywhere, as well as to people in countries like Africa without access to mains electricity or costly batteries. There is also a demand from aid agencies wishing to provide radios to people in emergency situations. These advantages are likely to overcome the disadvantages of having to wind up the spring, some noise from the gearbox and generator, and its relatively high price (about £60 when first introduced in the UK).

Baylis withdrew his patent in 1996 when earlier patents on clockwork-powered electrical devices were found, and other manufacturers started making clockwork radios (see Block 2 Section 4). To maintain its advantage the Freeplay range has continuously been improved with smaller, more efficient wind-up radios that no longer use clockwork mechanisms (Figure 80a). The manufacturer has since used the principle to diversify into other products such as wind-up torches (Figure 80b).

Compatibility. The styling of the original Freeplay was designed to appeal mainly to African users. New designs are intended to appeal mainly to Western consumers. The original Freeplay was compatible with values of sustainability – *ecological* (no pollution or waste from battery manufacture, consumption and disposal, or from generating electricity) as well as *social* and *economic* (providing employment for disabled workers in a developing country). The latter elements of sustainablity are less clear since the shift of manufacture to China.

Complexity. The concept of a wind-up radio is simple to understand and the product is easy to operate.

Observability. The original Freeplay radio is a fairly observable product in that it is distinctive looking and can be used almost anywhere. Subsequent designs have included visible features such as backup solar cells and transparent casings so users can see the clockwork mechanism (see the centre photograph on the front cover).

SAQ 15

How might the characteristics of the market, the innovator, regulation and government policy have helped or hindered the diffusion of the original Freeplay radio into the market?

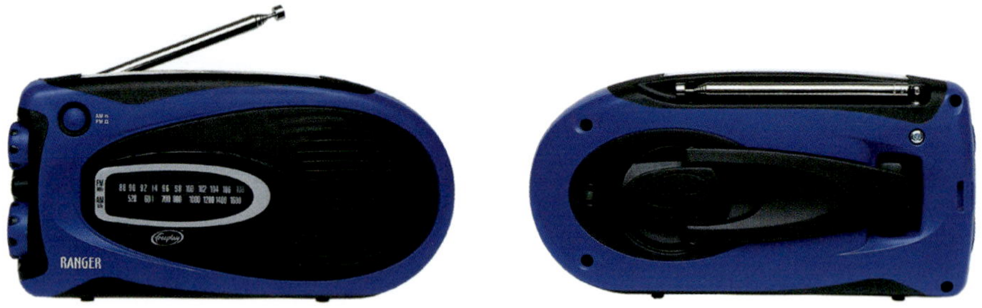

(a) Freeplay Ranger radio can run on wind-up power or rechargeable batteries fed by solar cells or a mains adapter

(b) Freeplay Sherpa wind-up and rechargeable battery torch

Figure 80

Environmental impacts of radios

We've looked at factors that enabled the orginal Freeplay radio to become a successful innovation in commercial and social terms. But there is another increasingly important measure of the success of a product or innovation – its impacts on the natural environment. If a product satisfies *social*, *economic* as well as *ecological* needs, it may be considered to be a *sustainable* design or innovation. How then does the Freeplay radio fare as an ecological product?

In Block 2, Section 4.2 you were introduced to life-cycle analysis (LCA) as a tool for assessing the total environmental impacts of a product.

Exercise 17 LCA for a clockwork radio

What is the basis of LCA? (Refer to Block 2 Section 4.2 if necessary.)

Considered as an ecological product, what phase of the total life cycle does a clockwork radio focus on? What environmental impacts might have been ignored?

Spend a few minutes making notes on these questions in your workfile before looking at my response below.

My response

LCA takes into account impacts arising from the extraction of *raw materials* from which a product is made, its *manufacture* and *distribution*, its *use* and *maintenance*, and its *disposal*, including any reuse or recycling.

By not requiring batteries or mains electricity, a clockwork radio focuses on the use and maintenance phase of the total life cycle. The other phases may not be considered at all in assessing its total impact.

A Dutch study has compared the life cycle environmental impacts of the original Freeplay radio with, among others, small Grundig and Philips battery powered portable radios (Stevels and Jansen, 1998).

The study found that the environmental impacts arising from the materials, manufacture and transport (from South Africa to the Netherlands) of the original Freeplay was three to five times that of the battery radios (imported from China). This was because of the Freeplay's greater size and weight, especially the weight of the steel spring. However, after about 18 months to three years of regular use the Freeplay had lower impacts than the battery radios because of the latter's consumption of batteries, even given the assumption in the LCA that the batteries were recycled (Figure 81).

The study concludes that human-powered devices can be, but are not necessarily, 'greener' than battery or mains-powered ones. It also said there was scope for improving the environmental performance of the original Freeplay clockwork radio, especially by reducing its size and weight. This has been done in later models (Figures 76 and 80(a)).

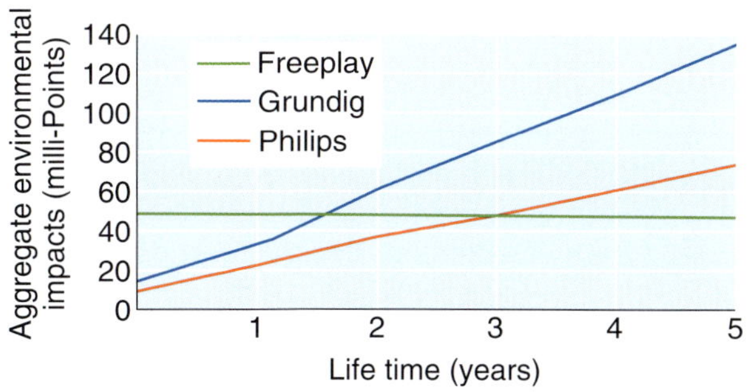

Figure 81 Aggregate environmental impacts of the original Freeplay radio and two small battery radios

The environmental impact of the manufacture, transport, use and end of life disposal of the original Freeplay clockwork radio and the Grundig and Philips battery radios over their life-time is shown. After 18 months to 3 years use, the Freeplay has lower total impacts (Stevels and Jansen, 1998, p. 55). The LCA employed the EcoIndicator95 method of measuring aggregate environmental impacts in milliPoints; this is the same as used in the *Eco-It* software on the DVD.

More generally, this case study shows that for a proper assessment of the ecological credentials of a product it is necessary to undertake a life-cycle analysis – guesswork can lead to incorrect conclusions.

Qualitative life-cycle analysis

The above study attempted to compare the life cycle impacts of similar types of product in *quantitative* terms. However, quantitative LCAs can be very complex and it is often only practicable to do a *qualitative* analysis. A useful qualitative LCA method is the so-called MET matrix, where

M = materials inputs and outputs;
E = energy inputs and outputs; and
T = toxic emissions or waste.

A blank MET matrix is shown in Table 3.

Table 3 MET matrix for qualitative life-cycle analysis

	Life cycle stage			
Environmental impact	Manufacture	Transport/ distribution	Use/ maintenance	Disposal/ recycling
Materials				
Energy				
Toxic emissions				

To conduct the analysis, you do the following. For a given product (e.g. the Freeplay radio) at each life cycle stage (manufacture, distribution, etc.) you enter qualitative assessments (stars, dots, question marks, etc.) to identify important environmental impacts in the areas of materials use, energy consumption and toxic emissions or waste. This process can be repeated for another product (e.g. a battery radio), as it's easier to compare products relative to each other using such qualitative judgements. For example, the Freeplay radio might be assessed ●●● (= high environmental impacts) under the materials stage of the life cycle because of its heavy construction, while a battery powered radio might get ★ (= fairly low impacts) because it uses less materials. At the use stage the Freeplay might be assessed ★★★ for materials, energy and toxic emissions because it needs no batteries or electricity, while the battery radios might be rated ●●. To make the assessments it is usually necessary to have some factual information, such as the comparative weight of the products, their energy consumption in use, etc. But clearly there is judgement and estimation involved in such an analysis. Nevertheless, a qualitative LCA provides a clearer picture of important environmental impacts than guesswork and can be very useful for choosing between concept designs and generating ecodesign improvements.

There is an opportunity to use the MET matrix to compare the original Freeplay and battery radios on the DVD. See the *Block 3 Supplement* for details. You may wish to attempt this DVD activity now or at the end of Section 6.

6.3.2 Electric-powered personal transport

Pedal cycles produce the lowest environmental impacts of any vehicle – a cyclist consumes only from 4 to 10 per cent of the energy per passenger kilometre of a car, bus or train – and utilises human energy for movement about five times more efficiently than walking. But the effort involved in riding cycles puts off many users. Hence there have been many attempts to design electric bicycles and other forms of ecological transport. However, none of these have so far had any significant success in most markets. Let's consider why.

The Sinclair 'C5'

One notorious failed innovation was the Sinclair 'C5' electrically assisted tricycle. Following a high profile launch in January 1985, the C5 sold only a fraction of the forecast 100,000 units per year. Production was discontinued after eight months when only about 5000 vehicles had been sold. Sinclair Vehicles went into receivership losing most of the £8.6 million that the project cost, including Sir Clive Sinclair's £7 million stake in the venture.

By risking his own money Sinclair had managed to overcome the obstacles to innovation. But could Sinclair have predicted the C5's commercial failure by considering the obstacles to diffusion you met in Section 6.2?

The C5's specification for maximum weight, speed, etc. was set by the 1983 Regulations on electrically assisted cycles that allowed it to be used by anyone over 14 years old without a licence, insurance or helmet (government *regulation*). Given this, did it offer a significant *relative advantage* over other modes of short distance transport? Sinclair felt that the C5 was so different that there was nothing to compare it with. However, that isn't the way that potential users were likely to view it. Once they realised that the C5 was not an electric car, but an electrically assisted tricycle that sometimes needed to be pedalled, most compared the C5 with its closest alternatives: the bicycle and the moped. Was this one-person vehicle, with a 15 mph (24 km/h) maximum speed, limited range and luggage capacity, costing nearly £400 (in 1985), worth buying? Other than for consumers attracted to the C5 for its sheer novelty (the *innovators*), the answer was likely to be 'no'. The C5 seemed to fail the first test of successful innovation.

Figure 82 The low riding position and lack of crash protection of the Sinclair C5 electrically assisted tricycle deterred many potential users

Second, was the C5 *compatible* with the values and skills of potential users? Sinclair's market research had produced favourable user reactions to a prototype. But after the launch it soon became apparent that there were very serious concerns about the safety of the C5, given its low driving position and lack of crash protection, especially since riders as young as 14 could use it in traffic without licence, helmet or insurance (Figure 82). And the performance of the C5 often did not live up to the advertising. In a society that values power and speed, it was probable that an unusual-looking electrically assisted tricycle, which many people had thought was as an electric car, wouldn't catch on.

Third, was the C5 too *complex* to learn to use? From my personal experience of trying a C5, it was quite easy to use, but for those who had not tried the vehicle the prospect of learning may have been a deterrent.

Finally *observability*. There is no doubt that the C5 was a highly observable innovation! But unusually this acted against it. This is because the C5 quickly attracted a disastrous image, and became the butt of numerous jokes. To ride a C5 seemed to invite public ridicule.

Sinclair was not alone in misjudging the demand for electrically assisted cycles. Many different designs have been developed and launched in the wake of the 1983 legislation, although none nearly as innovative or as expensively promoted as the C5. So far none have made a significant impact on the market, outside Japan and China (where 18 million electrically-assisted or powered cycles and scooters were made in 2007). Electrically assisted cycle design, constrained by regulation plus the low power/weight ratio inherent to battery vehicles, simply did not provide consumers with a big enough advantage over existing modes of transport to overcome the resistance to innovation.

It's easy to be wise in hindsight, but I think that you can see that testing the C5 concept against the obstacles to diffusion might have led Sinclair to abandon the project. Improved materials and battery technology have since allowed smaller and lighter electric cycles and scooters to be designed. Sinclair subsequently designed the 'Zike', a small electrically assisted bicycle, and other manufacturers have attempted to market power assisted cycles and scooters (Figure 83). However, the designs have either failed or only gained a limited niche market.

(a) Sinclair 'Zike' electrically assisted bicycle, launched 1992

The Zike cost £500, had an aluminium and plastics composite frame, and was powered by nickel cadmium batteries stored in the frame.

(b) Richard Seymour's 'Nexus' concept prototype, a butane-fueled motorcycle for city use

The design attempted to appeal to those who don't like motorcycles by altering the emotional response associated with them. (Photo courtesy of Seymour Powell Ltd. 1992.)

(c) 'City Bug' foldable electric scooter

This has a maximum speed of 22 km/h and a range 12–19 km (re-creation plc).

Figure 83 Power assisted personal transport vehicles

The 'Segway' human transporter

Let's now see what you make of another radical innovation – the Segway human transporter (Figure 84). Can you assess its likely market success from what you have learned?

(a) The 'Segway HT' human transporter launched in the US in December 2001

The 'i' Series model, priced about $5000 in 2008, has a top speed of 20 km/h and a range of 16 km for power assisted walking.

(b) The Segway 'i' Series in use on a pedestrian path

Note the small front panniers. (There is a larger luggage carrying model with additional large side panniers).

Figure 84

The Segway was conceived by an American inventor/designer/entrepreneur, Dean Kamen, who had made his fortune designing innovative medical equipment. A forerunner of the Segway was a self-balancing wheelchair called iBOT that enabled disabled people to climb stairs and kerbs. Kamen thought he could apply the idea of self-balancing transport in a device for people with full mobility. The basic idea behind the Segway was to provide a powered vehicle to enable people to travel short distances faster and further than they would walk, but without separating them from routes, such as paths or pavements, used by pedestrians.

The Segway is without doubt a marvel of design and technology, developed by a team of highly qualified engineers, designers and collaborating companies. It was first demonstrated publicly in December 2001. It is a hi-tech, two-wheel battery electric personal transporter that uses gyroscopes, tilt sensors, microprocessors and electric motors to keep balance. Lean forward and it starts, lean further and it goes faster. Twist a grip on the handle to turn. Straighten up and it draws to a halt; lean back and it reverses. There is no throttle or brake. Five gyroscopes and two tilt sensors continually monitor the orientation of the machine. These control the wheels and keep the rider balanced at the desired speed and direction.

The lighter 'i' Series model has a top speed of 20 km/h (12.5 mph, about 3–4 times faster than walking), a typical range of 16 km (10 miles) per battery charge, weighs 38 kg and costs about $5000 in the US (£4800 in the UK). There is also a luggage carrying model equipped with three panniers and an off-road 'x' series model. In the US and the UK the Segway is being sold online.

One British tester commented that riding the Segway was,

> fantastically enjoyable ... at its maximum the Segway feels terrifyingly fast, mainly because you are standing up eyeball to eyeball with pedestrians.

Another tester said,

> It is great fun ... you can go fast. It is very quiet. It isn't revolutionary but if they were asking maybe £1500, I think more people would buy them. The design is good; it also feels very good quality.
>
> (Booth, 2003, p. 12)

The Segway's 'footprint' (the ground space it occupies) at 48 × 64 cm is not much greater than an adult pedestrian and it can turn in its own space. It is thus claimed that the Segway will enable people to quickly and easily move and carry luggage wherever they can walk – in cities, campuses, theme parks, factories and warehouses, etc. It is further claimed that many Segway users, including postal and light delivery service providers, will not need cars for local travel especially in cities. Its developers say that if 10% of local car trips under 8 km (5 miles) were replaced by the Segway it would reduce fuel consumption by 8.6 billion litres and carbon dioxide emissions by 55 million tonnes per year in America alone (Segway, 2001–2). However, despite an attempt by MPs in 2008 to change regulations, to date the Segway cannot legally be used in the UK on public roads or paths. In the US while many states allow the Segway on public roads, a few allow it only on pavements or cycle paths, or ban its use in public areas.

Exercise 18 Successful innovation

On the basis of the above description, and any further information obtained as described in the *Block 3 Supplement*, assess whether the Segway is likely to succeed as an innovation. Use the obstacles to diffusion discussed in Section 6.2 to structure your assessment. (You could make a rough mock-up of the 'footprint' of the Segway in card or paper to stand on and try to imagine what it would be like to use.)

Spend 10–15 minutes on this exercise making notes in your workfile before looking at my response below.

My response

Characteristics of the innovation

Relative advantage: the Segway offers several unique features – a self-balancing, highly manoeuvrable personal powered transport device that occupies little space and seems great fun to ride. It could potentially be used where cycles cannot go, but it has no speed, weight or weather-protection advantages over conventional bicycles or electrically assisted cycles. Indeed, it seems to have significant disadvantages, notably its high price, the need for the user to stand and to carry heavy luggage on their back (at least for the non-luggage carrying model). It seems likely to be in competition with electric scooters (Figure 83(d)) and Segway imitators sold at a fraction of the price.

Compatibility: the Segway is very different in design to conventional bicycles and scooters and needs different skills to use, but might appeal to some users on the basis of its advanced technology and enjoyable ride. There have been objections from disabled people to Segway riders mixing with pedestrians.

Complexity: although in practice most people quickly learn to use the Segway, it might be perceived as difficult to learn and its technical systems would be complex to maintain and repair.

Observability: the novelty and distinctive design would attract some buyers, but the highly conspicuous appearance and potential for mixing with pedestrians might deter others.

Characteristics of the market

The Segway might appeal to non-cyclists, but it would need to become a highly fashionable 'fun' product in order to overcome consumer resistance. It may find niche markets for users such as postal delivery, warehousing, police, the military or in theme parks. Its success will depend crucially on what legislation or regulations apply to its use in different markets.

Characteristics of the innovators

The Segway was developed by a team of professional engineers and designers, assisted by several companies and launched using sophisticated marketing methods. This is certainly a factor in the Segway's favour.

Conclusion

It seems likely that, despite its highly sophisticated design and technology, for its price the Segway offers too few practical advantages to ordinary consumers over bicycles and existing electrically assisted cycles and scooters to succeed as an innovation. It may, however, find some commercial niche markets. At the time of writing (2008) the Segway has sold around 2000 units since its UK launch.

SAQ 16

Examples in this section show that ecological issues can be both a constraint on and a *stimulus* to creative design and innovation. Can you recall other examples from Blocks 1 and 2, and this block, of products in which ecological issues provided a stimulus to (or at least could have been a factor in) innovative design? On which environmental impact(s) did each example focus? (Look at the MET matrix in Section 6.3.1, Table 3.)

6.4 Conclusion

By now you should have learned that creativity is a vital, but is not the only factor, in the design, manufacture and introduction of a successful innovative product. You should have learned about theories of creativity, especially for creative concept design and invention. You should have learned that everyone can produce creative work and have attempted some exercises that make use of, and I hope have developed, your own creative design abilities. You will have the opportunity to practise those creative abilities in the Block 3 TMA by applying the approaches of experienced designers and where necessary using creative problem solving techniques.

If you haven't completed Exercise 8 at the end of Section 3 that tests your understanding of some of the theory covered in this block, you should do so now. You should also look back at the aims and learning outcomes at the beginning of the block and reread the section(s) or the key point summaries if you are unclear about any of the learning outcomes.

Try also to complete Section 6 of the *Modelling Workbook* before continuing to Block 4.

Key points of Section 6

Creativity is not enough to ensure a successful innovative product. Success requires that the *technical, financial* and *organisational obstacles to innovation* and the *obstacles to diffusion* are overcome.

Rogers' characteristics of an innovation that influence how rapidly it is likely to diffuse include: *relative advantage*; *compatibility*; *complexity* and *observability*. Other factors that affect diffusion are: the characteristics of the *market*, of the *innovator* and of any *regulations, legislation* and government *policies* concerning the innovation.

Rogers' characteristics can be used to explain, or predict, the rate at which, and the extent to which, an innovation diffuses into the market.

A quantitative or qualitative life-cycle analysis is required to assess, or compare, the total environmental impacts of product(s).

Ecological factors can be a constraint on or a stimulus to creative design and innovation.

(Learning outcomes 1.5, 2.4 and 3.3)

Answers to self-assessment questions

SAQ 1

Look back at Figures 86 and 87 in Block 1.

Are there any important differences between the design cycle models of designing in Block 1 and Figure 15 presented here?

Although the labels used differ, the models are essentially the same. They all show designing as a repeated cycle of generating and then evaluating/selecting ideas, concepts or solutions as the design develops. The model presented here (Figure 15) simply emphasises that each cycle often involves re-examining the problem or specification in the light of what was learned from the previous iteration (like Figure 87 in Block 1) .

SAQ 2

Define each of the terms in the following three groups. Then explain the relationships between the terms within each group:

(a) idea, concept design, technical principle, configuration; creativity;

Idea – the fundamental thought or notion for solving a problem.

Concept design – an embodiment of the idea in a sketch or model that has a technical principle and a configuration.

Technical principle – the scientific or technological basis on which a device or system operates.

Configuration – the arrangement of the parts of a design and how they are connected.

Creativity – a human activity that results in novel work that is perceived to be of value to individuals or society.

Relationships: *creativity* is required to conceive an *idea* and embody it in a *concept design* involving a particular *technical principle* and *configuration*.

(b) invention, patent, technical innovation;

Invention – an idea or concept for a new or improved device, product or process.

Patent – legal protection for an invention that is new, not obvious and capable of being applied in a practical device, product or process.

Technical innovation – the first commercial introduction or use of an invention.

Relationships: not all *inventions* are *patentable*, neither do all inventions become *technical innovations*.

(c) solution space, systematic design, heuristic design.

Solution space – a notional space within which lie all the possible solutions to a design problem.

Systematic design – a design process that encourages designers to methodically explore the problem and the solution space to find an optimum design.

Heuristic design – an approach normally used by experienced designers that enables them to home in on promising designs in the solution space on the basis of their knowledge and experience.

Relationships: *systematic design* and *heuristic design* are alternative strategies for exploring the *solution space*.

SAQ 3

What creative thinking process(es) are likely to have been involved in the following cases?

1 The concept design for a bridge based on the vertebrae of a dinosaur that its designers produced after a visit to the Natural History Museum in London.

Biological analogy

2 A designer using a pistol grip originally designed for guns in the design of other objects from power drills to cameras.

Adaptation

3 A nineteenth century designer's dissatisfaction with the uncomfortable saddles used on cycles of the time that led him to design a bicycle with a hammock like seat.

Constructive discontent; adaptation

4 The idea for an 'email pen' that enables the user to write an email, send it via a radio link to a small 'inkwell' shaped computer which transmits it to the Internet and on to the recipient (Figure 32).

Analogy, transfer.

SAQ 4

What types of modelling were used in the development of the first Cyclone cleaner? What models were used subsequently to develop new improved designs?

A principle proving model: a cardboard cyclone attached to an upright vacuum cleaner to test basic feasibility.

Development models: thousands of brass and aluminium models to develop optimum cyclone shapes.

Working prototypes to create and develop the novel hose configurations.

Subsequently, drawings and physical models to develop new designs, such as the first cylinder model, and mathematical/computer modelling to develop more efficient cyclone shapes and arrangements.

SAQ 5

Distinguish between the technical principle, the design configurations and the design features of the Cyclone cleaner.

Technical principle: use of cyclonic technology to collect dust.

Design configuration: double cyclones placed side by side and concentrically in the prototype and first production versions. Eight small 'root cyclones' in later models.

Design features: telescopic hose; colour schemes; transparent dust container.

SAQ 6

According to Dyson, how expert do you have to be in a particular technical field to produce worthwhile innovations?

At the initial concept stage it is best not to be *too* expert because the designer has to question established ideas. During development of the concept into something that works and can be economically manufactured it is usually necessary to become technically expert. However, becoming sufficiently expert is not very difficult when focused on a specific problem and area of knowledge.

SAQ 7

Why does Dyson consider that practical work with real physical objects is necessary for creative design and technical innovation? What other approaches are commonly used?

Practical work allows unexpected discoveries to be made through observation. Developing the technology of the Cyclone cleaner, for which Dyson did not have an adequate theory, required practical experimentation. The novel telescopic hose system was created through Dyson 'fiddling about' with a full-size prototype. Other inventors and designers work by drawing, and by mathematical modelling based on theory, as well as by practical experimentation. Had Dyson been more theoretically oriented he might have employed mathematical models sooner.

SAQ 8

On the basis of what you have learned so far, do you consider that Edison's observation: 'genius is 1% inspiration and 99% perspiration' is valid?

This block challenges the idea that creativity requires someone to have a special kind of mind that makes them a 'genius' – and thus suggests that Edison's observation is valid. Creative design is the result of thinking processes available to everyone, but undertaken by people with the knowledge, skill and motivation to produce a worthwhile new solution to a problem.

The basic idea on which a novel product is based may come from associative thinking in a flash of inspiration or stem from a designer's accumulated repertoire of knowledge and experience. The hard work that precedes and follows the conception of this idea makes up at least 99% of the time and effort involved.

SAQ 9

What was the 'problem frame' underlying the basic idea that led to the Strida concept design?

Sanders' problem frame was his dissatisfaction with the complexity and inelegance of existing folding bikes. He questioned why they should be designed – like the Brompton – to fold as small as possible. This leads to his basic idea that the bike could fold into a 'stick with wheels at one end' something like the Maclaren baby buggy.

SAQ 10

What methods did Sanders use to translate his basic idea into concept designs?

Sanders used a method of total 'immersion in the problem' involving sketching, literature surveys and reviewing existing solutions. (He might also have applied 'immersion' in the meaning used in Block 2 by riding existing folding bikes.)

SAQ 11

What was the source of the idea for the design of the bottom joint?

A *transfer* of ideas from similar situations, in this case a seat belt clasp.

SAQ 12

Why is sketching so important in Sanders' approach to creative design?

Sanders describes sketching as 'a dialogue with yourself' or 'visual brainstorming'. The aim is get as many ideas down on paper as possible in order to be able to 'clarify vague ideas in the head' and to move forward.

SAQ 13

What lessons about creative concept design processes can you identify from the Splashguard and backpack examples in Sections 4.1 and 4.2?

A systematic approach to concept design – problem exploration; specification; idea/concept generation; concept selection; concept development – was used in both projects.

An important part of problem exploration involves either identifying essential functions using a technique such as functional decomposition (Splashguard) or identifying key sub-problems from experience (backpack).

Ideas and concepts may be produced using a technique such as morphological analysis to generate alternative ways of meeting essential functions (Splashguard) or more informally through experience and group discussion (backpack).

Promising concepts for development may be selected using a technique such as a selection matrix (Splashguard) or by identifying which concept best resolves the sub-problems identified at the beginning of the project (backpack).

Within a systematic design approach, creative problem solving techniques can be helpful to learner designers (Splashguard), while experienced designers can rely on their experience to create concept designs (backpack).

SAQ 14

Identify the common factors in the management and design practices of the teams at IDEO and Philips Design shown on the DVD.

Both organisations tackle long term, systems design and shorter term product design projects.

In both organisations, autonomous teams of two to six designers are given the task of working on a design project in order to provide solutions for management or clients to evaluate.

Both design teams use associative thinking and visual brainstorming to generate ideas and concepts.

In both organisations 3D visual prototypes and/or principle proving models are made rapidly to evaluate, test and develop concepts.

SAQ 15

How might the characteristics of the market, the innovator, regulation and government policy have helped or hindered the diffusion of the original Freeplay radio into the market?

The *market*. Aid agencies assisted diffusion in developing countries. In Western markets a clockwork radio is likely to appeal to environmentally conscious, innovative consumers.

The *innovator*. Baylis is a very determined and publicity conscious innovator and that has helped him develop his ideas and promote his products. He has featured in many newspaper and magazine articles as well as appearing in documentaries, popular TV shows and advertisements.

Regulation and policy. A clockwork radio has a powerful spring that would be a severe hazard if released. This had to be contained to meet safety regulations. Otherwise the original radio was highly compatible with national and international regulations and policies on sustainable development.

SAQ 16

Examples in this section show that ecological issues can be both a constraint on and a *stimulus* to creative design and innovation. Can you recall other examples from Blocks 1 and 2, and this block, of products in which ecological issues provided a stimulus to (or at least could have been a factor in) innovative design? On which environmental impact(s) did each example focus? (Look at the MET matrix in Section 6.3.1, Table 3.)

Products in which ecological factors did, or could have, provided a stimulus to innovative design include:

Block 1, Section 6

- The chairs created to reduce the environmental impacts of the materials, energy and toxic emissions from manufacture – by designing with organic or recycled materials; to be durable or adaptable; and/or to avoid toxic substances.

Block 2, Section 4

- Kodak's single-use camera designed for component reuse or recycling to reduce the environmental impacts of manufacture and disposal.
- A coffee machine designed to minimise energy consumption in use.
- Office photocopiers designed to be leased, rather than purchased, to enable them to be returned to their supplier for refurbishment, upgrading or recycling, thus reducing the impacts of manufacture, use (via improved efficiency) and disposal.

- Designing shared services, e.g. car-sharing schemes, rather than selling products to individual consumers thus reducing the environmental impacts of manufacture (by efficient utilisation) and use (via fuel efficiency and good maintenance).

Block 3

- Wave power devices that minimise the impacts of energy generation and consumption (Figure 13).
- A multi-function printer/scanner/photocopier/fax that should reduce the impacts of materials in manufacture, of energy in use, and of disposal (Figure 29).
- The bicycles, folding bicycles, electrically assisted and alternative fuel vehicles that can reduce the environmental impacts of personal travel, especially energy and emissions from use.
- The human-powered battery charger (Figure 70(b)) to eliminate the need for electricity in use.
- The clockwork radios and torch (Figure 80) that reduce the materials etc. impacts of battery use and disposal.

References

Amabile, T.M. (1996) *Creativity in Context*, USA, Westview Press.

Arlidge, J. (2003) 'Father of invention', *The Observer*, 21 December.

Booth, M. (2003) 'The verdict: Segway', *The Independent*, 25 October, p. 12.

Burrows, M. (1993) 'Eureka!' (Interview by Barbara Myers), BBC Radio 4, 4 October.

Chick, A. (1997) 'The 'Freeplay' radio', *Journal of Sustainable Product Design*, Issue 1, April, pp. 53–56.

Christensen, C.M. (2000) *The Innovator's Dilemma*, 2nd edition, New York, HarperBusiness.

Cross, N. (1996) 'Winning by design: the methods of Gordon Murray, racing car designer', *Design Studies*, vol. 17, no. 1, pp. 91–107.

Cross, N. (1997) 'Descriptive models of creative design: application to an example', *Design Studies*, vol. 18, no. 4, October, pp. 427–440.

Cross, N. (2000) *Engineering Design Methods*, 3rd edition, Chichester, John Wiley.

Darke, J. (1979) 'The primary generator and the design process', *Design Studies*, vol. 1 no. 1, pp. 36–44.

de Bono, E. (1977) *Lateral Thinking. A Textbook of Creativity*, Harmondsworth, Penguin.

Demaid, A. (1995) 'The design of bicycle frames, in A. Demaid and J. de Witt, *Case Studies in Manufacturing with Advanced Materials, Volume 2*, Amsterdam and London, Elsevier.

Dunlop, J.B. (1922) *The History of the Pneumatic Tyre*, Alex Thom.

Dyson, J. (1987) 'Designer as Entrepeneur', transcript of talk to The Bath Design Conference 1987.

Dyson, J. (2001) *Against the Odds: an Autobiography*, (new edition), London, Texere Publishing.

Evamy, M. (1995) 'No socket required', *Design*, Winter 1995/96, pp. 50–53.

Fiell, C. and Fiell, J. (2001) *Designing the 21st Century*, Cologne, Taschen.

French, M. (1988) *Invention and Evolution: Design in Nature and Engineering*, Cambridge, Cambridge University Press.

Fuad-Luke, A. (2002) *The Eco-Design Handbook*, London, Thames and Hudson.

Geschka, H., von Reibnitz, U. and Storvik, K. (1982) 'Idea generation methods: creative solutions to business and technical problems', *Battelle Technical Inputs to Planning, Review no. 5*, Columbus Ohio, Battelle Memorial Institute.

Hickman, R.P. and Roos, M.J. (1982) 'Workmate', *CIPA Journal*, Chartered Institute of Patent Agents, July, pp. 424–57.

Hunt, M. (1991) 'Pots and pans, products and passions', *RSA Journal*, February, pp. 169–175.

Irvine, A. (1992) 'Mystery of form: architecture of Alvar Aalto', Exhibition review, *RSA Journal*, December, p. 60.

Lawson, B. (1994) *Design in Mind*, Oxford, Butterworth-Heinemann.

Lawson, B. (1990) *How Designers Think*, 2nd edition, London: Butterworth Architecture.

Lloyd, P. and Snelders, D. (2001) 'What was Philippe Starck thinking of?', in Lloyd, P. and Christiaans, H. (eds.) *Designing in Context*, Delft, The Netherlands, Delft University Press, pp. 240–255

Moulton, A.E. (1966) 'Design and technological innovation', paper given to the Design Congress 'Profit by Design'.

Moulton, A.E. (1979) 'Innovation', *Journal of the Royal Society of Arts*, December, pp. 31–44.

Moulton, A.E. (1997) 'A lifetime's experience of engineering innovations: successes and failures', *RSA Journal*, April 1997, pp. 10–13.

Myerson, J. (2001) *IDEO: Masters of Innovation*, London: Laurence King.

Nahum, A. (1988) *Alec Issigonis*, London, The Design Council.

National Advisory Committee on Creative and Cultural Education (1999) *All Our Futures: creativity, culture and education*, London, Dept. for Education and Science.

Open University (2001) Block 2 'Engineering by design', OU course T173 *Engineering the Future*, Milton Keynes, The Open University.

Otto, K. and Wood, K. (2001) *Product design: Techniques in Reverse Engineering and New Product Development*, New Jersey, Prentice-Hall.

Pugh, S. (1991) *Total Design: Integrated Methods for Successful Product Engineering*, Wokingham, Addison-Wesley.

QED (1995) 'Clockwork radio', BBC TV programme produced by John Groom, 8 August (updated 25 September 1996).

Ritchie, A. (2001) Transcript of interview with Adrian Demaid at Brompton Bicycles Ltd., May.

Rogers, E.M. (1983) *Diffusion of Innovations*, 3rd edition, New York, Free Press.

Roy, R. (1983) Block 2 'Bicycles: invention and innovation', OU course T263 *Design: Processes and products*, Milton Keynes, The Open University.

Sanders, M.A. (1985) 'The design of a new folding bicycle', Unpublished Masters Thesis, Royal College of Art/Imperial College of Science and Technology, London.

Segway (2001–2002) http://www.segway.com (accessed 6 March 2002)

Shahin, M.M.A., (1988) 'Application of a systematic design methodology: an engineering case study', *Design Studies*, vol. 9, no. 4, pp. 202–7.

Shah, J.J., Vargas-Hernandez, N. and Smith, S.M. (2003) 'Metrics for measuring ideation effectiveness', *Design Studies*, vol. 24, no. 2, pp. 111–134.

Starck, P. (2000) *Starck*, Cologne, Taschen.

Starley, W.K. (1898) 'The evolution of the bicycle', *Journal of the Society of Arts*, vol. 46, May 20, pp. 601–616.

Stevels, A. and Jansen, A.J. (1998) 'Renewable energy in portable radios: an environmental benchmarking study', *Journal of Sustainable Product Design*, issue 4, January, pp. 51–55.

Tovey, M., Porter, S. and Newman, R. (2003) 'Sketching, concept development and automotive design', *Design Studies*, vol. 24, no. 2, pp. 135–153.

Tyrell, R. (2002) 'Cavalier spirit', *Sunday Telegraph Magazine*, 9 June, pp. 18–19.

Ullman, D.G. (1992) *The Mechanical Design Process*, USA, McGraw-Hill.

Ulrich, K.T. and Eppinger, S.D. (2000) *Product Design and Development*, USA: Irwin McGraw-Hill.

Usher, A.P. (1954) *A History of Mechanical Inventions*, revised edition, Harvard College.

Van Gundy, A.B. (1988) *Techniques of Structured Problem Solving*, 2nd edition, New York, Van Nostrand Reinhold.

Van Dulken, S. (2002) *Inventing the 20th Century*, London, The British Library.

Wallas, G. (1926) *The Art of Thought*, New York, Harcourt.

Weisberg, R.W. (1993) *Creativity. Beyond the Myth of Genius*, New York, W.H. Freeman.

Whitfield, P.R. (1975) *Creativity in Industry*, Harmondsworth, Penguin.

Acknowledgements

Grateful acknowledgement is made to the following sources for permission to reproduce material within this book:

Illustrations

Figure 1: Biblioteca Ambrosiana, Milan; Figures 3, 46d, 51 and 52: MAS Design Products Ltd.; Figure 4: Philippe Starck, Agence Starck; Figures 5 and 36: Dyson; Figure 6: Mary Quant; Figure 8: Hoover Candy Group; Figure 9: Clare Jones; Figure 10: Ford Motor Company; Figure 11: Ulrich, K. and Eppinger, S., *Product Design and Development*, 2000. Copyright © 2000, 1995 by The McGraw-Hill Companies, Inc. All rights reserved. Reproduced with permission of The McGraw-Hill Companies; Figure 14: *Engineering Design Methods*, Nigel Cross. Copyright © 2000 by John Wiley & Sons Ltd. Reproduced by permission of John Wiley & Sons Limited; Figures 16 and 17: 'Design core: Conceptual Designs', Stuart Pugh, *Total Design*, Pearson Education Limited and Copyright © 1991 Addison-Wesley Publishers Ltd; Figure 18: Georgy Holden; Figure 25: Adapted from Lawson, B. (1990) *How Designers Think*, 2nd edition, Butterworth Architecture; Figures 26a, 30, 49b and c and 79a: Richard Hearne; Figure 37: Mike Levers; Figures 27, 35, 53, 69 and 83a: Robin Roy; Figure 28: Handspring; Figure 29: Hill & Knowlton; Figures 31b, 64 and 65: IDEO

Figure 32: Cambridge Consultants Ltd; Figure 33: Shah, J.J., et al, 'Metrics for measuring ideation effectiveness', *Design Studies*, vol. 24, No. 2, 2003, pp. 111-134, Reprinted with permission from Elsevier Science; Figure 34: Gordon Murray; Figure 38: © Getty Images; Figures 39-42, 45a and c: Alex Moulton Bicycles; Figure 46a: Micro Folding Cycle from Pashley Cycles; Figure 46b: riese und müller GmbH; Figures 46c, 47 and 57: © Mike Hessey, mike@whooper.demon.co.uk; Figure 46e: Bicycle Magazine; Figure 46f: Topham Picturepoint; Figure 48: *Patent No. 1 580 048 Improvement in or relating to folding bicycles,* 1980. Crown copyright materials is reproduced with the permission of the Controller of Her Majesty's Stationery Office; Figure 49a: Adrian Demaid; Figure 49d: Brompton Bicycle Ltd; Figure 50: Steve Garner; Figures 56 and 74: Raleigh UK Ltd; Figure 60: Adapted from Ullman, D.G. (1992) *The mechanical design process*, USA, McGraw-Hill; Figure 63: Adapted from *Design Studies*, vol. 18, no.4, Cross, N. 'Descriptive models of creative design application to an example', copyright 1997, with permission from Elsevier; Figure 66: Reprinted from *Design Studies*, Vol. 24, Tovey, M., Porter, S., Sketching, concept development and automotive design, p.147, copyright (2003), with permission from Elsevier; Figure 68: D. B. Jones; Figure 70a: Sarah Liddell; Figure 70b: Justin Floyd; Figure 72: Otto, K and Wood, K. (2001) *Product design: Techniques in reverse engineering and new product development*, Prentice-Hall; Figure 75: Honda; Figure 78: ETP Ltd; Figure 79b: Chick, A, 'The Freeplay radio', *Journal of Sustainable Design*, No. 1 April 1997, The Centre of Sustainable Design; Figure 80: The Freeplay Group; Figure 82: Times Newspapers Ltd; Figure 83b: Seymour Powell Ltd; Figure 83c: Solex Electric Bike Store; Figure 84: Segway™ LLC.

Every effort has been made to trace all copyright owners, but if any has been inadvertently overlooked, the publishers will be pleased to make the necessary arrangements at the first opportunity.

Course Team

Academic staff

Ken Baynes, External Assessor
Catherine Cooke, Author
Nigel Cross, Author
Chris Earl, Author
Steve Garner, Author and Course Chair
Georgy Holden, Author
Robin Roy, Author

Consultants

Mark Evans, Contributing Author, Workbook 1

Course managers

Andy Harding, Course Manager
Amber Thomas, Course Manager

Associate lecturers

Jenny Burke
Nick Jeffrey

Production staff

Tammy Alexander, Graphic Designer
Margaret Barnes, Course Secretary
Philippa Broadbent, Print Buyer
Jane Bromley, Interactive Media Designer
Michael Brown, Video Editor
Daphne Cross, Assistant Print Buyer
Tony Duggan, Learning Projects Manager
Bernie D'Souza, Course Secretary
Barbara Fraser, Picture Researcher
Phil Gauron, DVD Producer
Richard Hearne, Photographer
Katie Meade, Rights Executive
Jane Moore, Editor
Jonathon Owen, Graphic Artist
Alex Reid, Narrator (DVD)
Ekkehard Thumm, Media Project Manager
Howie Twiner, Graphic Artist
Robert Wood, Editor